Elsevier *Global Energy Policy and Economics* Series

European Electricity Systems
in Transition

Editor: Atle Midttun

Titles of related interest

Books

Handbook of Natural Resources and Energy Economics
Modern Power Station Practice *(12 Volumes, Major Reference Work)*

Journals *(free sample copy sent on request)*

Electric Power Systems Research
Energy Economics
Energy Policy
International Journal of Electrical Power and Energy Systems
Natural Resources Forum
Resource and Energy Economics
Resources Policy
Utilities Policy

For further details please contact your nearest Elsevier Regional Sales Office—see verso of main title page.

Elsevier *Global Energy Policy and Economics* Series

European Electricity Systems in Transition

A Comparative Analysis of Policy and Regulation in Western Europe

Edited by

ATLE MIDTTUN

ELSEVIER

Elsevier Science Ltd, The Boulevard, Langford Lane, Kidlington, Oxford, OX5 1GB, UK

Regional Sales Offices

EUROPE   Elsevier Science, PO Box 211, 1000 AE, Amsterdam, The Netherlands
              Tel: +31 20 485 3757 Fax: +31 20 485 3432
              Email: nlinfo-f@elsevier.nl

AMERICAS   Elsevier Science, 665 Avenue of the Americas, NY 10159-0945, USA
              Tel: +1 212 633 3730 Fax: +1 212 633 3680 Email: usinfo-f@elsevier.com

JAPAN   Elsevier Science Japan, 9-15 Higashi – Azabu 1-chome, Minato-ku,
              Tokyo 106, Japan
              Tel: +81 3 5561 5033 Fax: +81 3 5561 5047
              Email: kyf04035@niftyserve.or.jp

FAR EAST   Elsevier Science Pte Ltd
AND   No 1 Temesek Avenue
AUSTRALASIA   #17-01 Millenia Tower
(except Japan) Singapore 039192

Copyright © 1997 Elsevier Science Ltd

First Edition 1997

Library of Congress Cataloging in Publication Data

European electricity systems in transition: a comparative analysis of
   policy and regulation in Western Europe/editor, Atle Midttun. –
   1st edn
      p.   cm. – (Elsevier global energy policy and economics series)
   Includes index.
   ISBN 0–08–042994–7 (hardcover)
      1. Electric utilities – Government policy – Europe. 2. Electric
   utilities – Law and legislation–Europe. I. Midttun, Atle, 1952–.
   II. Series.
   HD9685.E82E943   1996
   333.79′32′094–dc20   96–44349
                                                                      CIP

British Library Cataloging in Publication Data

A catalogue record for this book is available from the British Library

ISBN 0 08 042994 7

Transferred to digital printing 2005
Printed and bound by Antony Rowe Ltd, Eastbourne

Contents

List of Contributors

Atle MIDTTUN is an Associate Professor at the Norwegian School of Management and Director of its Centre for Electricity Studies.

He has a PhD in economic sociology from Uppsala University, and has studied sociology, philosophy, history and economics as part of his masters degree at the University of Oslo.

His theoretical orientation is organisation theory and institutional economics, and much of his work applies this theoretical perspective to the energy and environmental fields. Some of his energy-related publications are: *The Politics of Energy Forecasting* (Oxford University Press), *Company Strategies and Societal Interests in Norwegian energy industry* (in Norwegian) (Vett or Viten). *A Comparative Study of Environmental Taxation in the Electricity Sector in the Nordic Countries* (Scandinavian Political Studies). *Strategic Challenges and Institutional Patterns: Nordic Heavy Industry's Adaptation to Market Oriented Electricity Reform* (Scandinavian Economic History Review).

Lutz MEZ is a Senior Associate Professor at the Department of Political Science and Vice Director of the Environmental Policy Research Unit at the Free University of Berlin.

He is a political scientist, expert in environmental and energy policy.

His major publications in this field are: *RWE—ein Riese mit Ausstrahlung* (Cologne: Kiepenheuer & Witsch 1996), *Energie-wirtschaft–Energiepolitik* (Script for Humboldt-University Berlin, 1994), *Die Energiesituation in der vormaligen DDR* (Berlin: edition sigma 1991), *Der Atomkonflikt, Reinbek: Rowohhlt 1981 Energiediskussion in Europa* (Villingen-Schwenningen: Neckar Verlag 1979FF).

Steve THOMAS is a Senior Fellow, Energy Programme at the Science Policy Research Unit of University of Sussex since 1979.

He has a BSc (honours) degree in Chemistry from the Bristol University. His research interests are in the structural change in the electricity industry, the structure and competitiveness of the heavy electrical industry and the economics of nuclear power.

His major works in this field are: *The Realities of Nuclear Power* (Cambridge University Press, 1988), *The World Market for Heavy Electrical Equipments* (Reed Publishing, 1990), *Electricity in Europe: Inside the Utilities* (Financial Times Business Information, 1992), and has been the major contributor to *The British Electricity Experiment* (ed. J. Surrey), Earthscan, 1996.

Frede HVELPLUND is an Associate Professor at the Institute for Development and Planning at Aalborg University in Denmark.

He has a Masters degree in international economics and a minor degree in social anthropology.

He has worked within the energy area since 1974 with special interests with connection between institutional conditions and technology development. He has published extensively on the Danish energy policy.

Some of his latest publications are: *Erneuerung der Energisysteme in den neuen Bundeslandernaber wie?* with Niels Winther Knudsen and Henrik Lund (Netzwerk Dezentrale EnergiNutzung, Potsdam, 1993). *Public regulation and technological change* with Henrik Lund (in Danish, Aalborg University, 1994). *Europaische Energiepolitik und Gruner New Deal–Vorschlage zur realisierung energiewirtschaftlicher Alternativen* (Institut für okologische Wirtschaftsforschung, Berlin, 1994). *Democracy and change* with Henrik Lund, Karl Emil Serup, Henning Mæng (in Danish, Aalborg University, 1995).

Rolf W. KUNNEKE is Assistant Professor at the Faculty of Public Administration and Public Policy at the University of Twente, The Netherlands.

He is an economist and specialized in the fields of industrial organisation and (neo- and new-) institutional economics. For years, his research interests have been focused on the restructuring of energy markets.

Some of his recent publications in this field are: (with M. J. Arentsen) *Economic organisation and liberalization of the electricity industry*, Energy Policy, vol. 24, no. 6, 1996., *Modelling welfare effects of a liberalization of the Dutch electricity market*, (with M. H. Voogt, Energy—The International Journal, forthcoming) *Marktwerking in de energysector* (with Arentsen, Manders, Steenge, Voogt, The Hague, 1996).

Maarten J. ARENTSEN is the Vice Director of the Center for Clean Technology and Environment Policy at the University of Twente, The Netherlands.

His research concentrates on regulation in regard to energy and environment.

His major publications are: *Radiation Policy in Netherlands* (Dietz F. J. and W. J. M. Heijmans (eds.)), *Environmental Policy in a Market Economy* (Wageningen, 1988), *Beleidsorganisatie en beleidsuirvoering. Een onderoek naar het stralenbeschermingsbeleid in Nederland, Engeland en Belgie* (diss., Enschede, 1991), and with Rolf Kunneke, *Economic organisation and liberalization of the electricity industry*, Energy Policy, vol. 24, no. 6, 1996.

Marcelo POPPE is a Senior Fellow at the CIRED—International Research Centre on Environment and Development, Paris, France.

He has graduated in Electrical Engineering from the Fededral University of Rio de Janeiro, Brazil and has post-graduated (DEA degree) in Energy Systems Economics by the University of Paris IX-Dauphine.

His present fields of interest include the institutional setting of energy sector, regulation of power markets and management of public services; energy efficiency, technological and organisational innovations in the energy field; energy planning, energy policies, environment and development issues; independent power production and renewables.

Some of his major recent works in this field are: *Renewables for power-policies for promotion of autonomous generation in Europe*, (CIRED Working Paper to LTI Project to DG XII-CE). *Production décentralisée d'électricité à partir des énergies renouvelables et de la cogénération: analyse des conditions de développement et d'achat on Europe*, together with D. Finon (IEPE) and S. Rego, Report to the French Ministry of Industry, IEPE, 155 Page, Grenoble, July 1995.

Lionel CAURET is a Ph-D candidate in economics at the CIRED—International Research Centre on Environment and Development.

His main fields of interest include the least cost planning, Integrated Resource Planning (IRP), Demand Side Management (DSM) and the French DSM ways, load curve modelling, new technical modelling of water-heaters and lighting, and particularities of the electricity grid in small islands (French Overseas Territories).

Some of his papers are: Cauret, L. and Adnot, J., *L'outre mer, des espaces electriques atypiques: synthese à l'occasion des cinquantenaires des DOM det d'EDF* (Revue de l'Energy n. 478—June 1996). Cauret, L. and Adnot, J. *Why Optimise an already efficient system? Overview of the French DSM approach* (19th International IAEE Conference, Budapest, Hungary—27–30 May, 1996). Cauret, L. and Adnot, J., *Planification intégrée de l'energie dans les DOM*, (1994 and 1995 Reports for ADEME—1994 and 1995).

Part One

Introduction

Chapter I
Restructuring Electricity Systems in Transition

LUTZ MEZ, ATLE MIDTTUN AND STEVE THOMAS

The fact that there is regulatory debate around the shaping of electricity systems is in no way a new phenomenon. The organisation of electricity systems since the late 19th century has had many stages and provoked several political debates.

I. The Early and Mature Structuration of the Electricity Systems

The first electricity systems were organised along decentralised lines. The large power losses of direct-current transmission necessitated small, locally situated power stations. Edison, the early pioneer of electricity, therefore aimed at an integrated organisation of all central stations, supplying lighting to users. His vision for the organisation of the electricity industry was therefore a large number of small, service-oriented utilities.

However, the industry gradually moved towards a larger integrated system. The first important invention to move electricity beyond the local level was the transformer. This made it possible to link urban centres to power stations situated far away, thermal to hydropower stations, and rural to urban areas. The transformer stimulated further technological developments. Alternating current became the dominant technology, and the turbine replaced the steam-engine. Over time, steam-turbine sizes increased a thousand-fold, as did the voltage in transmission systems. The efficiency of steam turbines increased by a factor of seven. These various economies of scale pushed down the real price of electricity over the course of the century, providing a basis for mass consumption and the emergence of national utilities.

The growing social importance of electricity triggered legislation and regulation. Shortly after the turn of the century, the new technological development of electricity and its perceived character of a natural monopoly led to public intervention, firstly through private franchised monopoly, and then gradually in many cases to full public owner-ship.

The dominant model for electricity-sector organisation was now char-acterised by the conceptualisation of electricity as a public infrastructure and part of the process of nation-building. It was a basic element in industrial policy and an important service to be made accessible to all consumers. In this period, large investments were therefore made in electricity systems, and subsidies were often made available to expand electrification to remote rural areas.

The post-World War II reconstruction put the electricity sector in focus as a major factor for modernisation, and again pushed the electricity-supply industry on to the political agenda of several European countries, strengthening the public ownership position. The develop-ment of nuclear and other large-scale technologies in many cases created a symbiotic relationship between utilities and electrotechnical suppliers, giving rise to some of the most powerful techno-industrial clusters in Europe. The power centre of these clusters varied. For several countries, the purchasing of turn-key power stations led to a run-down of utility design and construction departments, whereas utilities in other coun-tries maintained their own technological competences in this field.

The perception of electricity as a public infrastructure with natural monopoly characteristics, and the organisation of the sector into pub-licly owned or franchised institutional monopolies' led to a build-up of powerful sectoral configurations, dominantly operating as closed national systems. Coordination between these systems was undertaken on a voluntary basis, organised by sector associations like the UCPTE and NORDEL. Trade between them was largely a matter of marginal exchange of surplus to balance nationally independent production sys-tems, and the exchange prices were usually based on short-term mar-ginal costs. Depending on the resource base and national institutional traditions, some countries centralised the electricity system at the national level, whereas others anchored the electricity system organisa-tionally at the regional and local levels. In all cases, however, the man-dated public organisation or franchising agent had exclusive monopoly rights to supply customers located within its domain.

II. The European Variations

However, the commonality of the basic principles underlying the electricity supply industry did not prevent considerable diversity in the systems which existed up to the time of the current wave of restructuring. This diversity reflected factors such as political traditions and natural-resource endowments. The various European electricity supply industries differed in many important respects, for example, technology and fuel choice, ownership, and degree of vertical integration (see Table I.1). For example, The Netherlands stands out as the only case where the operation of the high-voltage transmission network was carried out by a company other than the dominant generator.

The German system (excluding the system in operation in the former GDR, which was run along very different lines) is particularly complex, and is best seen as eight separate but federated systems. Some were fully vertically integrated, while in others, separate local distribution companies operated. Some systems were dominated by coal, some by nuclear, and some were mixed. Public ownership was also at a number of levels, from the Länder which comprised the Federal Republic down to small local authorities.

However, despite this variety, the industries almost invariably operated under four shared assumptions which shaped the way the industry did business. First, all costs incurred could be recovered from consumers or tax-payers. Second, there was a close identity between utility policy and national government policy, regardless of the ownership of the utilities. Third, electricity was regarded not as a normal consumer good, but a service which should be available to all at affordable prices. Fourth, electricity supply systems were largely nationally self-contained and operated by companies which had electricity as their prime and usually their sole business.

These four assumptions had very powerful consequences for

Table I.1. The structure of European ESIs prior to liberalisation

	Technology and fuel choice	Ownership	Vertical integration
UK	Coal + nuclear	National public	Part
France	Nuclear	National public	Full
Norway	Hydro	National, regional and local public	Part
Netherlands	Gas and coal	Regional and local public	Part
Germany	Coal + nuclear	Regional and local public and private	Full and part
Sweden	Nuclear + hydro	National, regional and local public and private	Almost full
Denmark	Coal + gas	Regional and local public	Part
Finland	Nuclear + diverse	National, regional and local public and private	Almost full

planning and decision-making in the electricity supply industry. New investments were effectively underwritten by the consumer and therefore carried little economic risk to the utility despite the huge scale and technical complexity of much electricity supply industry plant. Electricity industry decisions on investment were justified on protecting the long-term interests of consumers and carried with them the implicit authority of government, and could not readily be deflected by third parties. Electricity industry decision-making used as its point of reference international electricity industry experience rather than general industrial experience. However, this international perspective was seldom used to compare the performance of national systems, and differences in consumer prices were generally assumed to be the consequence of differing national resource endowments rather than differing efficiencies. The international electricity supply industry can therefore be likened to a gentleman's club, where each member would abstain from invading the other's territory, and where members with common interests but different circumstances could meet to share experiences.

III. Recent Challenges to the Electricity Systems

The traditional electricity system has, since the early 1970s, been subject to increasing pressure in most industrialized countries, with dramatic implications for restructuring and change. The pressure comes from several sources, ranging from the rise of environmentalism, market forces and changes in the overall demand for electric energy, to technological innovation.

The pressure exerted by ecological problems has been increased since the beginning of the 1970s, making the electricity supply industry a subject to an ecologically motivated reevaluation. This process has involved demands for (1) clean air protection motivated by health hazards and later by forest damage, (2) resource conservation which appeared in the aftermath of the oil crisis to justify the objectives of energy saving, (3) landscape preservation in view of open-cast coal mining or large-scale hydroelectric power stations, (4) radiation protection, which became an issue in Europe especially after Chernobyl, and (5) climate protection, which in the late 1980s introduced new aspects of energy conservation to reduce CO_2 emissions.

The climatic risks represented by CO_2 set energy-saving as a prominent issue on the political agenda, marking a dramatic break with the traditional growth-oriented world view. Together with the stagnating electricity demand since the early 1980s, in several countries this made for a completely new outlook on energy consumption, shifting attention from investment and expansion towards resource optimisation within a given framework.

The pressure exerted on the electricity industry by ecological concerns and stagnating demand has been increased by simultaneous technology development. The economies of scale had for decades been taken as an economic premise for a massive expansion of the electricity industry. Over the last 10–15 years it has become increasingly clear that a new cost structure has arisen. New large-scale power stations became more expensive than existing ones and the rationality of bigger and bigger units was called into question by the possibility of co-generation, which attains its optimal effect in decentralized production units with short supply lines. Total energy units have been able to raise their overall efficiency rate considerably (to over 90%), and have thus become even more cost-efficient. In Denmark, The Netherlands, the United Kingdom and Germany a boom has developed in this sphere, and is now seriously challenging the market of the traditional electricity industry.

IV. Patterns of Change

The weakening of the prestige and power of the electricity supply industry following the challenges noted above has led to a progressive erosion over the past 15 years of the four assumptions that shaped the behaviour of the industry: the pass through of costs to consumers and tax-payers; the association of government and utility policy; the status of electricity as a universally available public good; and the self-contained nature of the industry.

At first this was a slow process, because the electricity supply industry had little reason to try to transform itself and few external commentators had the detailed knowledge and credibility to present alternatives to a system which, in its own terms, was reasonably effective, even if not perfect. However, as experience accumulated of operating electricity supply systems under different regimes, electricity supply industries were increasingly called on to justify their operations. Within a broader perspective, this process was strengthened by the increased influence of free-market economists (to whom centralised monopolies were anathema) in economic policy-making, and by the emergence of information technologies which greatly eased the practical problems of breaking up some of the monopoly powers. This wave of liberalism came on top of an already strong ecological and environmental critique.

The first chink in the monopoly armour was provided by the PURPA legislation in the USA. This was intended as a minor measure to ensure that small power sources and electricity from industrial co-generation schemes were utilised where it made economic sense. However, the high cost of the nuclear power plants then being completed gave these alternative power sources a very easy target to beat and, at the instigation of the regulatory authorities, a large volume of new generating plant was built by non-traditional generating companies. This appeared to demonstrate that companies other than traditional electric utilities could build generating plant efficiently, and that renewable technologies such as wind-power could make an effective contribution to generating systems.

The next piece of evidence came with the privatisation of the gas industry in Britain. This was carried out with no effective restructuring and became increasingly unpopular with consumers unhappy that privatisation involved no more than substituting a public sector monopoly for a private sector. This signalled that future utility re-organisations would have to pay more attention to the interests of consumers.

However, it was the privatisation of the British electricity supply industry that dramatically accelerated the process. A number of the central tenets of electricity supply system-planning and operation were broken by this. The monopoly in generation was broken and replaced by system driven by day-to-day competition; choice for final consumers was progressively introduced, and the national high-voltage grid was separated from the generation sector.

The processes known as privatisation or liberalisation are in fact complex multi-faceted activities sometimes involving a number of separate aspects. For example, the reform of the electricity supply industry in England and Wales included five discrete changes:

1.   a transfer of ownership from central government to the private sector—privatisation;
2.   corporate structural change involving the splitting up of some of the companies and change in the scope of business of others;
3.   a change of operation of the generation market from a monopoly to a market-driven system;
4.   the progressive introduction of competition to the market for supply to final consumers; and
5.   the introduction of a formalised regulatory regime at arm's length from government replacing the direct, control exerted by government ministers.

This structure opens up a range of new commercial possibilities for private-sector companies offering goods and services related to the electricity supply industry and is being enthusiastically promoted by powerful international interests.

The liberal path has since then been followed by Norway, in 1991, and more recently by Finland and Sweden, in 1995 and 1996. In all three cases, liberalisation took place without ownership transfer as in the British model. However, in the Norwegian case, with a more developed spot market, now available to all three Nordic countries, and with a far more decentralised industrial structure which now serves to balance the more concentrated Swedish and Finnish electricity industry in a common internal Nordic free-trade market.

However, other countries have responded to the challenges faced by the electricity supply industry outlined by refining and expanding the public service model. In the French case this refinement has primarily occurred as an internal reorientation of the national monopolist Electricité De France (EDF), which up until now has been reasonably successful in meeting the challenges. At the commercial level, it has been able to propose special deals for customers based on cut-off options during peak-load hours. At the technical level, EDF is improving its quality of electricity supply. At the financial level, EDF has significantly reduced its indebtedness, thus reducing its dependence on interest and foreign exchange-rate stability.

EDF has also been actively seeking international commercial engage-ments, with considerable investments and operation both in Europe and in third-world countries, and with extensive export to European neigh-bours on a commercial basis. In addition, EDF is also actively concerned with exploring potential for energy savings, and is explicitly planning to play a major part in demand-side management in France. However, EDF also sees potential for new consumption related to the environmen-tal interests in fields such as urban and industrial wastes, electric cars and other electricity based transports.

The Danish reactions to the new challenges have also been formu-lated within a public-service model, but a public service model with a more decentralised and pluralistic design. The Danish model has main-tained a non-competitive, partly consumer-owned electricity supply. However, there has been an opening for the introduction of new wind and combined heat and power technologies on favourable terms, within an overall energy-planning framework, headed by central political authorities attentive to the political debate. This dual orientation of Danish energy policy has, on the one hand, fostered diversity and cre-ated pluralism both in technology and modes of organisation. On the other hand, it has resulted in some unresolved competitive tensions, which are currently open to political debate.

The Dutch system can be seen as a balance between a public service and a commercial model, with negotiated environmental agreements between industry and political authorities. The reform moves, spelt out within a public-service oriented model in 1989, have not yet resulted in a stable institutional setting but seems to be moving towards a market system of production and trade in electricity.

Germany still maintains its federal monoply model, with nine regional mixed public–private companies enjoying largely exclusive supply rights. The model has, however, been exposed to frequent challenges, both from German competition authorities and from industrial consumers and some municipal companies. The challenges have nevertheless all been successfully fought off, and market competition is thus kept away, at least for the time being. Government interests in using the electricity sector to finance a costly and non-competitive coal industry and local municipalities' interests in collecting large concession fees coincide with the electricity industry's interests in maintaining monopoly positions, and this grand coalition has proven hard to break up. On the environmental side, however, the German electricity industry has made extensive investments, responding to perhaps the most articulate environmental political forces in Europe.

The diversity of national models is reflected at the European level, where battles between liberalist and public-service factions have resulted in a rather vaguely defined coexistence based on subsidiarity, and some small steps in the direction of more liberal trade.

V. Conceptual Diversity

There seems to be a parallel between the complexity of the European restructuring of the electricity industry in practice and the complexity of regulatory styles, models and theory. The diversity of national regulatory styles, and models reflect the different political economies of Europe. The Anglo-American liberal tradition is confronted with a French tradition of 'regulation', a German tradition of 'Ordnungspolitik', a Scandinavian tradition of pragmatic state involvement and a Dutch tradition of negotiated agreement. Secondly, there are a number of theoretical disciplines, including economic theory, organisation theory and political science theory, focusing on different sides of the regulatory problem and recommending different remedies.

Within each of these disciplines there are again different schools with varying approaches to regulation. The Austrian approach, with its emphasis on dynamic growth processes and innovation, thus challenges the rationality assumptions found in much of the neoclassical thinking, and tends to lead away from a static optimisation towards a dynamic growth perspective. The formal approach in Weberian rational bureaucracy-inspired organisation theory is fundamentally challenged by the orientation towards informal social processes found in the human-relations and organisational-culture schools. In political science, the idealised concept of the democratic constitutional politics has again been challenged by, for instance, interest-group theory, which conceptualises state policy-making as undertaken not by a coherent actor but as an outcome of competing interests.

European electricity regulation is therefore likely to remain not only politically and institutionally but also theoretically a question of multiple positions and models, where the only general dominant conceptualisation is one of path dependency, or historical conditioning of systems. This again raises a fundamental theoretical (and practical) problem of integration. If there are large collective gains to be harvested from a more extensive European electricity trade, then some principles for coordination are needed across the boundaries of institutional diversity.

The European debate on regulation is now addressing exactly this point: how to integrate systems under institutional pluralism. Conceptual models such as negotiated third-party access and single-buyer arrangements have been presented so as to allow each system to preserve elements of their national institutional preferences. However, as we show in chapter IX on 'Electricity Policy in the European Union', no good theoretical solution has yet been found to integrate them under common trade.

However, the institutional diversity of European electricity regulation is not necessarily going to last forever. As a counter-hypothesis to the path-dependency argument for sustained European institutional and regulatory pluralism, one may take up the idea of institutional isomorphism, which has become an important argument within organisation theory. The institutional isomorphism hypothesis argues that organisations tend to model themselves on other relevant organisations as part of a mimetic processes in situations of uncertainty where goals and means are unclear, or that political influence and the need for political legitimacy drives organisations to similar solutions. In addition, the institutional isomorphism literature stresses the role of normative pressure arising from professionalisation as another factor leading to similar organisational design. This line of argument leads us to expect a European convergence rather than upheld diversity as the national electricity systems start relating more closely through international trade.

VI. The Structure of the Book

The main part of the book—chapters III–VIII—comprises national case studies of the development of regulatory and organisational responses to the recent challenges to electricity systems. These chapters give fairly rich descriptions of the restructuration processes and outcomes. Starting with a presentation of the pre-reform structure, the country studies review major developments of industrial structure, organisational reforms, technological conditions and political legitimacy issues. A separate chapter (chapter IX) is also devoted to the regulatory policy development of the European Union, where the focus is mainly on the political negotiations and institutional conceptualisation of a common regime.

Underpinning the empirical discussion of national electricity regulation, the book also conveys a more or less implicit debate over regulatory styles and approaches. A brief overview of some of the major positions in economic and organisation theory is given in the following chapter on 'Regulation Beyond Market and Hierarchy', and each of the national cases more or less explicitly conveys descriptions of national regulatory models and regulatory style.

Chapters X and XI draw up a comparative analysis of regulatory models and styles and regulatory policy, respectively. The comparison of regulatory models summarises and typologises the regulatory and organisational approaches and reflects on the correlation between regulatory models and system performance. Chapter X draws together the regulatory experiences, reflecting back on the theoretical issues spelled out in chapter II. Chapter XI draws up a theoretical basis for politological comparison of national regulatory approaches. On this basis it discusses how national policy traditions have shaped regulatory models and regulatory responses to the new challenges.

In a concluding chapter, the book presents an outlook on the future European scene for electricity in particular and the closely related energy and environmental development more broadly.

Acknowledgements

We would, finally, like to thank the Norwegian Research Council for financial support to the co-ordination and the comparative analysis necessary to accomplish this book. Together with the systematic work of Ishwar Chander as an editorial assistant, this support has been essential for getting the book together.

Chapter II
Regulation Beyond Market and Hierarchy: an Excursion into Regulation Theory

ATLE MIDTTUN

I. Introduction

European electricity regulation is characterised by a diversity of models ranging from politically controlled hierarchies to market-oriented decentralisation. The variety of regulatory models imply considerable differences on a number of crucial dimensions, including goals, ownership, grid access, market concentration, openness to international trade, etc. Together with variations in national resource endowments, regulatory diversity has produced a corresponding diversity of prices, tariff structures and taxation regimes.

From a regulation theory point of view, this situation could be problematic. If one assumes that there is *one* optimal model, then the diversity of regulatory models and outcomes imply that European countries are suffering welfare losses as a result of erroneous regulation. But is such an assumption warranted? Does European regulatory diversity represent a deviation from the one optimal path, or is it a rational response to a more complex regulatory situation?

In this chapter a brief overview is given of some of the debate on industrial regulation, both in microeconomics and in organisational theory, which reveals that there is hardly any basis for a broad consensus on an *a priori* optimal path. A general theory of regulation covering both intra- and inter-organisational/market issues does not yet exist. Furthermore, new economic growth theory has highlighted the need to move beyond simple resource allocation and to start including dynamic learning processes into regulatory analysis. In addition, strong theoretical grounds can be given for relativising regulatory theory to specific national, regional and/or sectoral institutional conditions. No decisive theoretical grounds can therefore, in our opinion, be given *a priori* to reject the French hierarchical model, the British regulated oligopoly model, the Norwegian decentralised competitive market model, the Dutch negotiated consensus model, or the German industrial network model.

This does not imply, however, that we retreat to complete pluralism. Our claim is that the regulatory task, because of its theoretical complexity, is a tedious one of evaluating specific outcomes of existing and potential regulatory approaches in relationship to each other within each special national, social and cultural context.

Because of the complementarity as well as substitutability between markets and hierarchies, regulation theory should be intimately concerned with both economic and organisational factors and the interplay between them, as it indeed often has been. However, because of disciplinary specialisation, organisational elements have often been severely underplayed and unduly simplified in economic regulation theory, and likewise, market elements have often been underdeveloped in organisation theory. We shall try to break out of this one-sidedness by presenting some reflections on modern regulation theory while drawing on both traditions.

The parallel discussion of organisational and economic regulation is made easier by what we see as a parallel trend in the refinement of both theoretical perspectives. Within both economic and organisational regulation theory, it is possible to construct a development from simplistic, formalist, static, efficiency-oriented regulation towards a broader, complex, dynamic, growth-focussed regulation with a more procedural heuristic orientation. We shall discuss this development in four steps. Section II concentrates on control and static efficiency-oriented regulation. Section III discusses organisational and economic refinements of this position through what could be termed an enlargement of the regulatory menu. Section IV discusses regulation in a dynamic growth perspective, where economic and organisational regulatory instruments must be tailored to creative innovation. In Section V, we discuss regulation in its institutional context, and take up the issue of applying regulation to varying national styles and institutional conditions, but also of institutional isomorphism across national boundaries. Section VI focuses briefly on application.

II. Control and Static Efficiency

II.A. *The early public finance tradition*

Two main thrusts in early post-war economic regulation theory were (1) A belief in regulation through competitive markets as a default option with reference to idealised resource optimisation under free trade conditions, generalised to a whole market system in general equilibrium theory, and (2) a belief in regulation through public hierarchical governance under certain specified conditions. These conditions all concern so-called 'market failure', where simple free-trade competition would break down.

This theoretical discussion was analytically underpinned by a static efficiency perspective, where regulation was formulated as a resource-allocation problem which could be handled mathematically, and where the solution could be extracted through mathematical analysis. This analytical universe opened up the possibility of reaching precise and clear-cut solutions to the regulatory issues.

Besides the general focus on market-based resource allocation, early economic regulation theory, as expressed in the public finance tradition, was concerned with drawing up the demarcation criteria between market-based and hierarchically-based regulation, the underlying assumption being that perfect hierarchical governance was available within the public sector whenever competitive markets 'failed'.

II.B. The implicit organisational assumptions

In its underlying organisational assumption, the early economic regulation theory made an implicit alliance with early organisation theory, and more specifically with Weber's (1964) ideal model of bureaucracy, which in many ways constitutes the organisational parallel to the ideal type of free-trade market. Hierarchical governance according to this model rests on a combination of an impersonal authority structure, a hierarchy of offices in a career system of specified spheres of competence, free selection based on achievements in accordance with specified rules, remuneration in terms of money based on clear contracts, discipline, and control in the conduct of office.

Weber sets the tone for a general thrust in organisation theory that Scott (1981) has termed the 'rational systems' approach to organisations. At the core of this perspective is the view that organisations are seen as actions performed by purposeful and co-ordinated agents with goals providing unambiguous criteria for selecting among alternative activities.

II.C. The demarcation criteria

The demarcation criteria between market-based and hierarchically-based regulation in the early public finance literature include both consumption-side and production-side elements (Samuelson, 1954; Musgrave and Peacock, 1967; Lane, 1993). The first of the consumption-side arguments is the inexcludability argument, meaning that the good cannot be 'fenced in' from collective consumption. A typical example here would be a lighthouse. The second consumption-side argument is the public welfare argument stating that it is desirable to allow open access to a good even though it can be 'fenced in' if the marginal cost of adding another consumer is extremely small.

On the production side, the argument of scale advantages is central (Musgrave, 1959). The argument here is that if a given consumer segment can be more efficiently served by one producer than by many producers, then producer competition is impossible to maintain. Given the greater efficiency and hence the competitive advantage of the large producer, a free-trade market will degenerate into monopoly, and there will be no endogenous market incentives to evade monopoly profits.

II.D. Asymmetry between economic and organisational assumptions

A first and rather obvious critique of the early public finance position is one of asymmetry between organisational and economic assumptions. By concentrating solely on conditions for market failure, and implicitly assuming public hierarchical success in cases where regulation through market mechanisms failed, the public finance perspective was obviously making unrealistic assumptions, quite out of tune even with contemporary research in organisation theory.

An extensive literature on bureaucratic failure constitutes a direct parallel to the market-failure literature within the public finance tradition. Some of the main themes in this critique have been that bureaucratic or formal organisation is vulnerable to displacement of means and ends (Merton, 1957; Selznick, 1949), that there is a tension between formal structures and informal processes (Mayo, 1945; Rothlisberger and Dickson, 1939; Perrow, 1970), that formal organisation encourages opportunistic behaviour (Niskanen, 1971; Parkinson, 1957; Starbuck, 1965; Downs, 1967), and that formal organisation may in some cases erode to chaos (March and Olsen, 1976).

If we take this more developed organisational perspective into account, we are left with a more intriguing set of regulatory choices (Table II.1), where the criteria for demarcation between market-based and organisational regulation are more unclear. Whether to go for market-based or hierarchically-based regulation must now depend on a careful evaluation of success and failure of both regulatory alternatives.

Table II.1. Market, hierarchy, success and failure

	Public hierarchy success	Public hierarchy failure
Private market success	1	2
Private market failure	3	4

In addition to the complexity problem involved in adding on a more developed organisation-theory perspective, the early public finance theory has also been more endogenously criticised for lack of clarity in its demarcation criteria. Critics have argued that non-excludability, welfare advantages of joint consumption and scale advantages were not clearly distinguishable as indicators of market-based or hierarchical governance. Closer analysis has revealed that very few goods are in fact characterised by inexcludability or non-subtractability in a strict sense. To save the theory's empirical relevance, a less absolute delineation, based on economic criteria, therefore had to be introduced. However, these economic criteria in turn make the theory much less sharply applicable.

II.E. *The market–hierarchy position reconsidered*

The joint effect of bringing in the organisational dimension by adding the discussion of organisational failure to the discussion of market failure, and of taking the critique of the demarcation criteria into consideration, is to complicate the regulatory task severely. The extreme cases of full market failure and full hierarchical success, as well as full market success and full hierarchical failure, remain simple to diagnose. However, these are hardly the empirically relevant cases. A more realistic task is one of choosing between a whole range of more or less successful market and hierarchical solutions and of carefully evaluating their pros and cons.

III. Enlarging the Regulatory Menu

The breakdown of the simplistic assumptions of the early public finance theory—that market failure could always be compensated by a successful public hierarchy, and that there were clear demarcation criteria for selection among the two options—has led to a search for a broader set of regulatory means. The core question within economically based regulation theory has changed from when to substitute unsuccessful market governance with hierarchy to how to modify unsuccessful market governance with supplementary means. This approach, which we will call external incentive-based regulation, implies working through the market actors by supplementing the market with additional control and incentive structures, rather than to substitute them with administered regimes. This approach has been widely applied in general market regulation, for instance by internalising environmental externalities. By adding supplementary external incentives, for instance through taxes or tradeable permits, it has been possible to add new and more specific regulatory targets without abolishing the basic market mechanism. Adding supplementary external incentives has even made it possible to maintain elements of market competition in so-called natural monopolies.

The orientation towards supplementary incentives in economics has its parallel in organisation theory. Regulatory refinement within organisational theory can be described as an expansion through development of internal incentives, where efficiency is furthered through incentives built into the organisational structure and processes, as opposed to the external incentives approach in economics. In the organisational refinement approach, the internal organisation of the market actors themselves are the targets of regulation.

III.A. External incentive-based regulation

The task of economic external incentive-based regulation is to move the adaptation as close as possible to the social optimum by creating an incentive environment which leads the firm to seek such solutions while acting in its own interest. The challenge of supplementing the simple general market mechanism has been most pressing in the case of so-called natural monopolies. For this purpose, several external incentive models have been devised, including rate-of-return regulation, price-cap regulation, yardstick regulation, and menu-of-contracts regulation.

Rate-of-return regulation is a regime which allows a firm an 'acceptable' return on invested capital, but which does not allow a firm to gain a surplus beyond this. Within these boundaries, the firm is allowed to choose technology and input factors, and to decide production quantum and prices. Rate-of-return regulation has been widely applied in many countries, but perhaps most prominently in the USA, where it is applied to the electricity sector.

In spite of its popularity, rate-of-return regulation has important weaknesses. The fact that the incentive effects of this mode of regulation are related only to the capital base constitutes a basic problem (Averch and Johnson, 1962), as one of the goals of regulation is to secure efficiency for the firm as a whole. The Averch–Johnson effect illustrates that rate-of-return regulation has a potential weakness when it comes to information. The information held by the public regulator may not suffice to prevent the private hierarchy from exploiting its customers.

Price-cap regulation is a rather new addition to the incentive-based regulation repertoire. It has enjoyed great popularity in Great Britain, where it is applied to gas and electricity regulation following the infrastructure reforms of the late 1980s and 1990s. The basic characteristic of this mode of regulation is that the regulatory authorities specify a maximum price for the product. This maximum price can also be adjustable with a general factor, for instance referring to the factor prices of the input factors of the regulated industry in order to promote productivity. The British formula $RPI - x$, for instance, has an inbuilt productivity factor x which pushes the firm to increase is productivity.

Price-cap regulation contains a mixture of factors which are exogenous to the firm, such as the development of the relevant price index and factors which the firm can affect, namely productivity. The problem still remains, however, of how to calibrate both the productivity demands and the price cap at the starting point, where the regulator is heavily dependent on insight into the firm's operations, and where the firm has an incentive not to reveal information.

Yardstick regulation builds on comparisons between firms, or between firms and a model firm for regulation. The economic results of a firm are compared to those of other comparable firms, and the firm is regulated on this basis. If the regulation is not based on comparison with a norm-model, this method presupposes that a sufficient number of comparable firms exist to allow the method to be applied on an empirical basis. Implementation of yardstick regulation presumes reasonable homogeneity within the group. If this is not the case, statistical methods may be applied to sort the group into subcategories where yardstick competition may be applied. This presupposes, however, that there are objective criteria available for sorting the companies into respective groups.

A more sophisticated method of yardstick competition has been devised in the so-called data envelopment analysis method. This method focuses on the performance of entities in a multidimensional analysis of, for instance, capital and labour productivity. For each dimension, as well as for the whole set of dimensions, the analysis allows us to specify the productivity 'front' and to compare each entity to it (Norman and Stoker, 1991).

Even given the assumptions of large numbers, which in one sense defies the idea of monopoly, yardstick competition may have problems with sustainability over time, given the actors' strategic behaviour in response to expected future outcomes of regulation. If revealing efficiency in the first round could result in tougher restrictions in the next round of regulation, then companies will be reluctant to reveal information which yardstick competition presumes.

With its roots in modern information economics, Laffont and Tirole (1993) have formulated a regulatory regime where the firm is faced with a menu of regulation contracts, where it may choose among different combinations of cost-sharing and rewards. Contracts with low rewards will typically be associated with the firm wishing to take on a small share of future variations in cost. On the other hand, the firm may choose to take on contract alternatives with high rewards, but which subject the firm to taking responsibility for cost variations. The first type of regulation contracts will closely resemble rate-of-return based regulation, while the other will be closer to price-cap models.

The incentive mechanisms described in this theory have not been extensively tested; it is therefore difficult to judge its merits adequately. As for the other incentive-based regulatory approaches, a critical problem for the 'menu of regulation contracts' approach is information. The specification of the contractual regime presumes considerable knowledge on the part of the regulator about commercial detail, which may be difficult to obtain from the firms. Another critical question is whether such a regulatory approach is sustainable over time, given an actor's strategic behaviour in response to the expected future outcomes of regulation.

III.B. *Internal incentive-based regulation*

As already mentioned, the orientation towards supplementary incentives in economics has its parallel in organisation theory where we may speak of a strategy of regulatory refinement through 'internal' incentives.

The move towards more advanced models of internal incentives has taken several paths. One path has been to develop formal internal incentive structures which further desired goals at both societal and enterprise levels. Another path has sought to supplement formal organisation with additional informal incentive structures. This tradition has sought to evade organisational failure stemming from incongruence between formal and informal structures within the organisation. In addition, organisational theorists have come to focus more and more on the organisations' external relations as a major design component.

Scott (1981) summarises the main lines in the development of modern organisation theory—which provides the basis for internal, incentive-based, regulation—in a two-dimensional model. He combines along the one axis a shift from formal to organic or natural systems, and along the other axis a shift from closed to open systems (Table II.2).

The first, closed, formal rational model (type I) is identical to the idealised Weberian hierarchy discussed in the previous section. This was the dominant theoretical model for analysing organisations up to the late 1930s, and as such, hardly belongs to this section on 'enlarging the regulatory menu'. Nevertheless a mainstream endeavour within organisation theory and administrative practice has been to refine and develop further the principles of hierarchical regulation presented in the core of the Weberian analysis. This refinement still occupies a central position in organisation theory and should therefore also be counted among the instruments of modern regulation. Some of the aims of this approach have been to:

1. clarify the goal structure of the organisation;
2. strengthen routines for goal implementation;
3. strengthen the unity of leadership and clarify authority relations; and
4. tailor organisational units and subunits to tasks, so that homogeneous or related activities take place within the same organisational unit.

The formal and closed character of this system of analysis was instrumental in allowing strong normative conclusions. However, the organisational theorists within this model conceived of organisations as tools designed to achieve preset ends, and underplayed perturbations and opportunities posed by connections to a wider environment. In all of these models, including contributions from Weber (1964), Taylor (1911) and Fayol (1949), the variety and uncertainty associated with an organisation's openness to its environment was assumed or explained away.

The regulation-design implications of this approach have been to strengthen and refine the formal organisation in order to achieve better and more precise governance. This approach is clearly most applicable to a regulatory task characterised by stability and predictability, where

Table II.2. Developments in organisational design

	Formal rational model	Natural/organic model
Closed system	I	II
Open system	III	IV

Adapted from Scott (1981).

external goals are set and the regulatory task is to optimise the organisation with respect to fulfilling those goals.

The formal rational model was, according to Scott (1981), succeeded by the closed natural systems model in the late 1930s and early 1940s. One of the hallmarks of the natural systems approach was that it took the informal dimension into account more consistently. Given the assumption that informal behaviour patterns were more crucial for the capacity of the organisation to operate successfully than formal characteristics, the key to organisational and regulatory efficiency was seen to lie in various improvements of the social and psychological milieu. The human-relation school, and later research following in its wake such as that by Trist *et al.* (1963), Crozier (1964) and Clark (1960), thus came to emphasise factors like personal motivation and satisfaction, group integration and identification and democratic leadership as core instruments for improved organisational and regulatory efficiency.

The regulation-design implications of this tradition is obviously to inform regulatory analysts about the informal structures and interpersonal systems of power, and of the impact of such elements on the formal structure. Regulatory models are explicitly or implicitly built on a set of behavioural assumptions which, if they are not present, or undermined by informal behavioural patterns, may seriously distort the regulatory outcome. The promise of the natural systems approach is to target not only the formal but also the informal parts of the organisation towards the regulatory tasks.

A third approach to organisational improvement, which Scott (1981) calls the rational open-systems approach, has been to consider external challenges more closely. This approach came as a reaction to the latent assumption in many of the internally focused studies that organisations largely controlled their own destinies. In many cases the assumption that the regulatory agency or system has exclusive control within its domain is clearly unreasonable, and the rational open-systems approach recognises this by incorporating external conditional variables into the regulatory analysis.

The open systems literature has focused on a large variety of external conditional variables, including size, technology and uncertainty, and has been concerned with how organisations or regulatory systems may respond rationally by choosing adequate formal structures which take the external conditional variables into account. Scott (1981) finds this approach exemplified by the work of Udy (1959), Woodward (1965) Pugh *et al.* (1969) and Blay (1970). However, the element of rational selection of formal structure in response to organisational environments is more explicitly represented in the work of design theorists like

Galbraith (1973) and Swinth (1974), but is also characteristic of the work of Thompson (1967).

The implication for regulatory design is that the struggle to develop efficient regulatory structures must relate to specific types of environments; in other words, that choice of the regulatory model is context-dependent. The rationalist assumption of this position is that the external conditioning variables are known and that adequate regulatory responses can be designed through formal organisational means.

The open natural systems approach to regulation maintains the focus on external conditions but abandons the strong rationality assumption of the formal, rational approach. Scott (1981) sees this position represented by the work of such theorists as Hickson *et al.* (1971), March and Olsen (1976), Meyer and Rowan (1977), and Pfeffer and Salancik (1978). These models resemble the rational open-systems approach in so far as they place great emphasis on the importance of external factors in determining organisational behaviour. However, the assumption that organisations are in a position to react rationally to meet these challenges is subject to debate.

Some organisational theorists go so far as to argue that the whole question of rational adaptation was irrelevant. The population ecology perspective thus argues for environmental determinism (Hannan and Freeman, 1977). However, more moderate theories reintroduce some degree of internal organisational discretion. Resource-dependency theory (Pfeffer and Salancik, 1978) or theory focusing on loosely coupled relationships between internal and external factors (Weick, 1980) are examples of this.

Resource-dependency theory sees organisations and regulatory systems as dependent upon the outside world for essential tangible resources. It focuses on the adaptation of the regulatory system to its external environment with respect to power and control: if the regulatory organisation has only weak control, it will have to compromise extensively with its external environment. If it has more control, it can pursue stronger regulatory policies.

The idea of loose coupling implies that environmental selection of regulatory strategies may take place over a very long period of time, or that there may be several niche possibilities that allow the existence of several organisational responses to the same regulation problem. This introduces a certain degree of uncertainty into the regulatory analysis, and indicates that regulatory responses may in part be process-dependent. Various regulatory solutions may thereby stabilise around several equilibrium points, dependening on the chosen trajectory.

At the extreme end-point of the regulatory challenge, under open natural-systems conditions, March and Olsen (1976) show that when organisations or regulatory regimes operate in environments in which goals are ambiguous and technologies are unclear, there may be little room for concerted organisational action to meet environmental challenges in a rational way. In such situations, regulation may deteriorate to a garbage-can process, where regulatory outcomes are products of more or less arbitrary couplings between actors, problems and solutions.

Even though the garbage-can model is an extreme case, the regulation-design implications of the open natural-systems situation are complex. External and internal complexity may make it difficult to specify, *a priori*, operational regulatory goals. Because such situations imply learning and reconceptualising of regulatory strategies, open natural-systems situations point beyond a static optimisation paradigm towards more dynamic growth-oriented regulatory thinking.

III.C. *Bridges between markets and hierarchies*

The extended regulatory menu provided by external and internal incentive modifications of the market and hierarchy ideal types provides opportunities for a tighter fit between regulatory problems and regulatory design. However, it also faces regulatory analysis with more complex choices. With a large set of modified market and modified hierarchy models, the demarcation between the two basic approaches to regulation becomes far less clear.

Rather than two distinct regulatory models, we may talk about two core ideal types, where each ideal type may be modified by a set of auxiliary elements (such as introducing external incentive elements in a markets and internal incentive elements into hierarchies). The 'domains' of market- and hierarchy-based regulation can thus be seen as a set of partially overlapping spheres, with the ideal type in the core and modified variants based on advanced incentive or co-ordinating mechanisms or other auxiliary elements further out in the periphery (Fig. II.1). Presumably, the further out from the core one moves, the more costly the regulation becomes, because of the complexity of administrating the auxiliary mechanisms. Thus we would expect that hierarchical approaches in the core of the market domain (for instance under ideal free-trade conditions) and market-based approaches in the core of the organisational domain (for instance under situations with natural monopoly and/or extensive problems of collective action) to be relatively costly compared to the other regulatory form.

The complexity of delineating the boundaries more sharply between the two approaches, however, takes us beyond the specification of simple demarcation criteria, to address the need for a more general

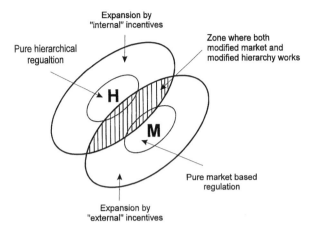

Fig. II.1. Regulatory approaches and domains.

bridging theory. Such a bridging theory would have to indicate an optimal choice of incentives/authority mechanisms for any regulatory problem, assuming that for both market-based and authority-based systems there is an opportunity to move out from the simple core through the use of advanced incentive or co-ordinating mechanisms, to more mixed modes of regulation. In the difficult overlapping grey zone, where both modified market-based and modified hierarchy-based regulation is possible, the bridging theory would have to indicate which modification is better.

Elements of such a bridge may be found in several theoretical traditions. Certain notions from the transaction-cost theory (Coase, 1937; Williamson, 1975) are clearly relevant. This theory suggests that the organisational path to regulation should be taken when the costs of measuring service outputs are high, and when we are concerned with investments in asset-specific skills and capital. In organisation theory, Stinchcombe (1984) has argued for contractual complexity as a criterion for deciding how to choose between market-based external incentive systems or authority-based organisational approaches to regulation. Under rather simplistic contractual conditions, the contractual set-up and the incentive regime could be heavily market-based. Under more complex contractual conditions, the regime would more closely approximate hierarchical conditions.

The similarity of the models of Stinchcombe (1984) and Williamson (1975) indicates a trend of parallelism between organisational and economic theory which could be promising in terms of creating a bridging theory of regulation (Barney and Ouchi, 1986). However, the possibility of solving the bridging problem between economic and organisation theory by the approaches of Williamson (1975) and Stinchcombe (1984) remains to be more extensively explored.

IV. Regulation for Dynamic Growth

Besides creating the need for an integrating bridging theory, the extended regulatory menu in general and the open natural systems model in particular point beyond static efficiency towards a dynamic growth perspective on regulation. By dynamic growth-oriented regulation, we imply regulation oriented towards dynamic evolution based on technological and organisational innovation, as opposed on the marginalist focus on price formation and equilibrium.

A theoretical basis for dynamic growth-oriented regulation can be found in economic theory, as well as within business strategy and organisation theory. Common to many of these theories is that they complicate the task of industrial regulation, as it can no longer target a future optimal state but must relate to a dynamically shifting sequence of optima resulting from new technological and organisational knowledge. Furthermore, adding realistic assumptions of cognitive limitation, these optima cannot possibly be fully defined. The dynamic growth-oriented perspective on regulation therefore also has methodological implications: under dynamic growth, process orientation and heuristic analysis takes the place held by mathematical optimisation under static efficiency.

As for static efficiency-oriented regulation, economically-based regulation is based more on rivalry, while hierarchically-based regulation puts stronger emphasis on co-ordination. However, the difference is less pronounced than under static regulation as both rivalry and co-ordination take on a different meaning under dynamic growth (Table II.3). Whereas rivalry under static, resource-based optimisation is ideally pushed to its extreme limit under free trade, rivalry under dynamic growth may take on a more limited form, and in certain cases

Table III.3. Regulation through rivalry and co-ordination under static optimisation and dynamic growth

	Dynamic growth	Static optimisation
Rivalry	Semi-competitive industrial clusters	Free trade regulation
Co-ordination	Learning-oriented open natural systems	Bureaucratically organised planned economy

remains only latent. On the other hand, co-ordination under hierarchical regulation might be looser and less monolithically controlled than under the bureaucratic ideal type.

The reason for this blurring of the earlier ideal types is that innovation and learning, which are vital to the dynamic growth process, may demand a pooling of industrial resources beyond the boundaries of one firm, and thus prescribe semi-competitive industrial clusters rather than free trade. Similarly, under co-ordinated governance, innovative processes may demand more open learning-oriented regulation than under the bureaucratic ideal type.

IV.A. Dynamic growth-oriented regulation through rivalry

Taking regulation by market rivalry as the dominant regulatory mode, it is recognised within the so-called new growth theories in economics that learning must partly be treated as a group activity. It then follows that even though market rivalry is the baseline model, dissipation of knowledge outside of market transactions must take place in order to achieve dynamic growth. Mainstream neo-classical economics therefore recognises that competitively based regulation may have to be supplemented with more co-operative mechanisms. It is, however, the Austrian tradition that has paid most attention to dynamic growth. In this tradition, the ability to further dynamic innovation and industrial restructuration is considered to be far more important for economic growth and welfare than marginalistic resource optimisation. In Schumpeter's (1943) perspective, economics should therefore be more concerned with disequilibrium and creative destruction than with marginalistic equilibrium analysis.

With their rejection of the equilibrium concept and concentration on a dynamic process perspective, the Austrian analysis is in essence entirely incompatible with any static understanding of economic activity. In fact, because of the focus on innovation and limited knowledge, the concept of process, in the Austrian understanding, establishes a fundamental 'indeterminateness' of economic activity which defies any concept of equilibrium and thus of optimality (Ioannides, 1992).

This acknowledgement, for many Austrians, leads to rather restrictive position on regulation. To the Austrians, the process perspective and the derived 'indeterminateness' assumption imply that the market systems constitute an entirely self-regulating order. Thus its functioning does not require any supra-economic regulation, except for the institutional arrangements of the liberal state. What could perhaps be defined as an Austrian-inspired regulatory position would be a policy of market-opening, where political authorities would see it as their task to tear down or abolish barriers to market entry, but where it would be up to commercial forces to utilise the opportunity space that this regulation creates

IV.A.a. Business strategy

Recognising, with the Austrians, that competition is dynamic and evolving, the business strategy literature shares much of the Austrian concern with innovation and learning processes. In the words of Porter (1990):

> Instead of simply maximizing within fixed constraints, the question is how firms can gain competitive advantage from changing constraints. Instead of only deploying a fixed pool of factors of production, a more important issue is how firms and nations improve the quality of factors, raise the productivity with which they are utilized and create new ones.

At the company level, business strategists point out that these targets can often be more easily reached through strategic alliances than through competition (Hax and Majluf, 1991). These alliances may include a number of institutional forms, all sharing the common objective of eliminating or significantly reducing confrontation among competitiors, suppliers, customers, potential new entrants and substitute producers.

At the industry level, business strategists are somewhat more open to an active role for governmental regulation than at least the hard-core Austrians like Hayek and von Mises. However, the mostly US-based business-strategy literature agrees with the Austrian position, that government regulation or intervention should be indirect as governments do not control national competitive advantage, but can only influence it.

In the business strategy perspective, governments can shape and influence the context and the institutional structure surrounding firms, as well as the inputs they draw upon, through measures including:

1. close collaborative efforts in R&D and specialised research institutions focused on industry clusters to create a match between science and technology policy and patterns of competitive advantage in the nation's industry;
2. signalling and targeting to direct a critical mass of societal and company resources in a common direction; and
3. measures to secure long-term investment: the fact that competitive advantage often is created over a longer period with concerted action implies that considerable attention must be directed at long-term investments and collective organisation, where government regulation has an important role.

The challenge of regulating a dynamic innovative economy in business strategy terms involves shaping a new role for government, beyond the minimalist liberal state. How far government should intervene is debated. Some strategists, like Porter (1990), are against strong direct interventionist regulation. State regulation in his perspective should remain indirect. This nevertheless leaves government with room for competent facilitation and market-friendly support. However, European business strategy analysts like Doz (1986) point out that industrial regulation often implies more tightly knit strategic interfaces between governments and companies, including such measures as government ownership and long-term protected contracts.

IV.B. Dynamic growth-oriented regulation through co-ordination

Hierarchical co-ordination for dynamic growth shares many characteristics with Austrian and business strategy rivalry-based regulation. As in the two latter cases, there is a recognition within natural and open systems-based organisation theory that co-ordination must make room for complexity and variability of the individual component parts. It is recognised that the system boundaries may be amorphous, and that the assignment of actors or actions to either the organisation or the environment may vary, depending on what aspect of system functioning is under consideration.

As Scott (1981) points out, open systems co-ordination does not simply blur the more conventional views of the structural features of organisation: it shifts attention from structure to process. Also here, the affinity with the Austrian position is evident. Co-ordination in this perspective therefore becomes process management, where enacting, selecting and retaining processes (Weick, 1969) becomes a core concern. In dynamic growth-oriented co-ordination, the process view is taken not only of the internal operations of the organisation but of the organisation itself, persisting over time (Scott, 1981). In order to cope with dynamic growth, the organisation itself must, in other words, be a dynamic system (Leavitt *et al.*, 1973). Open-systems based co-ordination stresses the reciprocal ties that bind and interrelate the organisation with those elements that surround and penetrate it (Scott, 1981). Pursuing this to the very limit, co-ordination for dynamic growth becomes a system of 'regulated improvisation' or generative rules which represent the internalisation by actors of past experience (Bourdieu and Passeron, 1977) with reference to a basic set of organisational tasks.

The shift from static bureaucratic control models to co-ordination through dynamic open systems also has methodological implications which are parallel to the transition from static to dynamic optimisation in economics. From analytical deductive reasoning under the bureaucratic hierarchy, organisational analysis under dynamic open systems tends to work heuristically, from analogies.

IV.C. *Convergence towards network-based negotiated regulation*

The similarities between the open natural systems model in organisation theory and the Austrian model of dynamic economic growth indicate a convergence towards negotiated network-based regulation. The modified rivalry under the dynamic growth model may, for instance, include extensive co-ordination between firms and governments in long term development agreements. Similarly, modified co-ordination under the open natural systems model may include a loosening up of direct control, amoebic adaptation of organisational boundaries, and inclusion of internal quasi-competition within the firm, implying a transition of governance to underlying negotiations. Network-based regulation is, according to Powell (1981) characterised by:

1. a normative basis of complementary strengths, as opposed to contracts and property rights for markets and employment relationships for hierarchies;
2. relational means of communication as opposed to prices for markets and routines for hierarchies;
3. norms of reciprocity and reputational concerns as methods of conflict resolution, as opposed to haggling and resorting to courts for markets and administrative fiat for hierarchies;
4. an open-ended mutual benefit climate, as opposed to a climate of precision and/or suspicion for markets and a formal bureaucratic climate for hierarchies; and
5. interdependent actor preferences or choices, as opposed to independence for markets and dependence for hierarchies.

A network-based conceptualisation of both firms and markets is indeed gaining momentum both in organisation theory and in business strategy. Aoki *et al.*'s (1990) concept of the firm as a nexus of internal and external treaties is a case in point. Similarly, Håkanson *et al.* (1989) have for a long time described relations between large multinational industries, customers and suppliers as having very much a network character. Such networks serve important functions in terms of innovation, efficiency, stability and control. The most important implication of the existence of relationships is that companies cannot be regarded as independent units able to chose counterparts at any time, but rather as units interlocked with each other, constituting highly specialized and complex structures.

V. Path Dependency, National Styles and Institutional Isomorphism

The discussion of dynamic growth-oriented regulation has left us with a seemingly broad choice of regulatory options where definite regulatory strategies cannot be pointed out on an *a priori* theoretical basis, but must remain heuristically sketched. Not only are both co-ordinated and rivalry-based strategies only open to heuristic process-specification in the case of dynamic growth, but there is also a considerable overlap between the two approaches which allows for several regulatory choices.

In this way, regulatory closure can only be achieved with considerable inductive input. To this input belongs, of course, the personal values and preferences of decision-makers, which, since they are idiosyncratic, shall not be discussed here. However, there is also a more general factor of industrial and cultural traditions that also form predispositions for regulatory choices and also for regulatory success, which can be discussed at a general level.

V.B. Path dependency and national styles

National styles of industrial decision-making and broader historical traditions form contexts, and are opportunities for industrial and regulatory development through what can be called path dependency. A central element in the path-dependency perspective is that industrial systems cannot develop independently of previous events (David, 1993). Local positive loops serve to propagate traditional patterns into future strategic decisions. This implies a development with several equilibrium points, where small events at one point in time may play an important role for future development by determining the course of a long-term development.

Given the theoretical openness and the inductive dependency, especially of dynamic growth-oriented regulation theory, path dependency can play a major role in determining regulatory strategy. Self-reinforcing processes of path-specific values and choices and the accompanying build-up of path-dependent regulatory competencies and resources may in fact drive two countries or regions into very different regulatory regimes for a given industrial sector, both with success. A number of researchers have pointed out that different countries and sectors have had commercial success with widely different developing models. On the one hand, the American car industry at the beginning of this century is brought in to illustrate great success with large-scale industrial organisation. On the other hand, Italian manufacturing industry is brought in to illustrate the success of flexible specialisation based on traditional family and village networks (Piore and Sabel, 1984).

The general path-dependency argument is given a more specific interpretation in a large socioeconomic literature which discusses national styles in industrial organisation under several labels such as business systems (Whitley, 1992), social systems of production (Campbell *et al.*, 1991) and modes of capitalist organisation (Orru, 1994). The essence of this literature is again that industrial development proceeds differently in different countries, as national industrial 'milieus' draw on specific traditions and competence in their national surroundings. Differences in industrial and industrial policy traditions can in this way be seen as producing three widely different models of 'capitalist' development: (1) competitive capitalism in the USA and Great Britain, (2) alliance capitalism in Germany and Japan, and (3) state-directed capitalism in France. All of these models, in certain respects, have their advantages.

Sensitivity to national styles and cultural conditions obviously adds yet another element to the regulatory menu. On the one hand, national predispositions, competencies and values can be brought in to create inductive closure in the face of analytical complexity. On the other hand, this may become yet another factor in complicating regulatory analysis

V.B. Institutional isomorphism

However, national differences are challenged by international learning, co-operation and/or dominance. This constitutes forces towards cross-national harmonisation, or what Di Maggio and Powell (1991) have termed 'institutional isomorphism'. Di Maggio and Powell (1991) apply this idea to organisations, but in our opinion it may equally well be applied to regulatory regimes. Di Maggio and Powell (1991) point out three mechanisms through which institutional isomorphic change occurs: mimetic, coercive and normative isomorphism.

Changes in regulatory regimes may occur as mimetic processes, where changes in relevant reference nations act as a signal to change, perhaps in response to uncertainty. When regulatory technologies are poorly understood, when goals are ambiguous, or when the environment creates symbolic uncertainty, states may model their regulatory regimes on other states. In this perspective, the regulatory regime of one country tends to be modelled on regimes in other countries which are perceived to be more legitimate or more successful.

Coercive isomorphism stems political influence and the need for political legitimacy acting as a driving force for institutional isomorphism. The strong formal and informal pressures exerted by the EU Commission for liberalisation of the energy and industry sectors in the late 1980s and early 1990s may have been felt as a force or a persuasion to undertake national reform in a liberal direction. The strong attempts by the EU to create a common legal environment conducive to liberal reform, had it succeeded, might have created a compelling force for liberalisation.

In addition, isomorphism may be closely associated with normative pressure arising from professionalisation, in those cases where the professions have or adopt strong regulatory regime preferences. As noted by Perrow (1974), since professionalisation tends to create a pool of almost interchangeable individuals who occupy similar positions across a range of organisations and even nations, professional similarities may override variations in tradition and control that might otherwise shape regulatory behaviour.

VI. Conclusions and Applications

VI.A. Control, growth and complexity

In our discussions of the development of regulation theory, within both market- and hierarchy-oriented thinking, we have noticed a shift from control- to growth-oriented analysis. The emphasis on control in much of the early literature stems from the need to set rules so as to limit economic activity within decent boundaries (environmental protection, quality restrictions, distributive justice, etc.). The emphasis on control has also been motivated by the need to maintain competitive rivalry within the economic approach, or to maintain perfect authoritative control within the administrative approach, under the assumption that such rivalry/authoritative rule respectively furthers efficiency.

The concern for dynamic growth in much of the later literature stems from the recognition that material welfare is fundamentally not only dependent on static efficiency but also on creative innovation. Regulation must therefore, besides the traditional efficiency concern, also be deeply concerned with dynamic growth in order to stimulate technological innovation, experimentation with new organizational forms, and learning and qualification of the 'human capital'.

The complexity of regulatory analysis, within the broader dynamic growth-oriented perspective and the idiosyncracies of national styles, commands some respectful constraints on drawing too bold theoretical conclusions. Both theoretical and methodological arguments have been listed to indicate that our regulatory redesign of complex organised industrial systems can only at best hope to represent improvements or second-best solutions, and never a dynamically changing and elusive optimum. At a synthetic level, therefore, regulatory design of dynamically developing industrial systems must largely remain a heuristic process-oriented exercise, with considerable scope for experimentation.

VI.B. Decomposition for regaining simplicity

The above conclusion does not preclude, however, that more stringent analysis may be partially applied to regulatory sub-issues which are sufficiently delineated to allow local static optimality conclusions. The so-called principle of decomposition can be utilised to break down complex industrial systems into partial regulatory tasks, to which more specific regulatory principles can be applied. A so-called 'value chain' analysis of industrial systems previously considered to be so-called 'natural monopolies' has, for instance, revealed that a large set of productive activities in electricity, gas and telecommunication systems can in fact be traded through market systems, while the natural monopoly element, which makes for a different regulatory design, only constitutes a limited component.

In a value-chain perspective, an electricity system, for instance, can be conceptualised as a set of activities ranging from generation through wholesale, high-voltage transport, low-voltage distribution and retailing to consumption (Fig. II.2), where only parts of the high-voltage transport and low-voltage distribution can be said to have elements of natural monopoly. In this case, control-oriented regulation may be applied only to selected parts, while the rest of the system may be targeted by dynamic growth-oriented regulation.

However, the principle of decomposition and partial analysis needs to be handled with care, and must constantly be iterated against the larger system context. Dynamic technological development may affect the regulatory challenges, for instance the natural monopoly character of a given function. Satellite communication is, for instance, changing basic premises for the earlier grid-based telecommunication monopolies, and allows for a transferral of new sections of the telecommunication systems from monopoly-based to rivalry-oriented regulation. Specification of partial regulatory designs to elements of the industrial system based on a partial analysis must therefore always remain temporary in the light of larger system developments which call for new industrial configurations and a more holistic regulatory reconceptualisation.

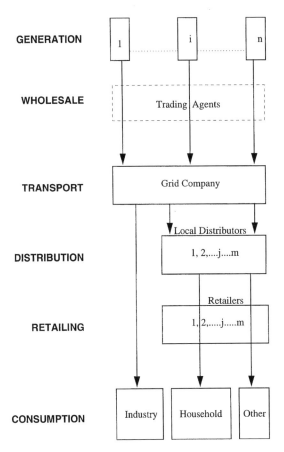

Fig. II.2 The value chain of electricity.

In spite of the possibility of partial and temporary static optimisation, regulatory designs may never be finalised, but must remain open to constant challenge and reconsideration. Part of that challenge will come from the dynamics of industrial development, as new technologies and new industrial strategies create new regulatory problems. Part of the challenge, however, comes from regulation theory itself, as economic and organisational analysis develop more and more refined menus of regulatory approaches. In this development, one may expect a close coupling of organisational and economic analysis, as economic incentives and efficiency will have to remain major concerns for organisational solutions, and as market-based regulation must always presume organisational elements both at the micro/firm and the macro/regulator levels.

Literature

Aoki, Masahiko, Gustafsson, B. O. and Williamson, Oliver E. (1990) *The Firm as a Nexus of Treaties*, Sage, London.

Averch, H. and Johnson, L. L. (1962) Behavior of the firm under regulatory constraint, *American Economic Review*.

Barney, Jay B. and Ouchi, William G. (1986) *Organizational Economics*, Joseey-Bass, London.

Baumol, W. J., Bailey, E. E. and Willig, R. D. (1977) Weak invisible hand theorems on the sustainability of multiproduct natural monopoly, *American Economic Review*, Vol. 67, p. 355.

Bourdieu, Pierre, and Passeron, Jean-Claude (1977) *Reproduction in Education, Society and Culture*, Sage, Beverly Hills, CA.

Campbell, John L., Hollingsworth, Roger and Lindberg, Leon N. (1991) *Governance of the American Economy*, Cambridge University Press, New York.

Clark, James V. (1960) Motivation in work groups. A tentative view, *Human Organization*, No. 4.

Coase, Ronald (1937) The nature of the firm, *Economica NS*, Vol. 4, pp. 386–405.

Crozier, Michl (1964) *The Bureaucratic Phenomenon*, University of Chicago Press, Chicago, IL.

David, P. A. (1993) Path dependence and predictability in dynamic systems with local network externalities: a paradigm for historical economics. In Foray, Dominique and Freeman, Christopher (eds) *Technology and the Wealth of Nations*, Pinter, London.

Di Maggio, Paul and Powell, Walter W. (1991) The iron cage revisited: institutional isomorphism and collective rationality in organizatanalysisional fields. In Di Maggio, Paul and Powell, Walter W. (eds) *The New Institutionalism in Organizational*, University of Chicago Press, Chicago, IL.

Downs, A. (1967) *An Economic theory of Democracy*, Harper and Row, New York, NY.

Doz, Yves (1986) *Strategic Management in Multinational Companies*, Pergamon, Oxford.

Fayol, Henri (1949) *General and Industrial Management*, Pitman, London.

Galbraith, Johen Kenneth *Designing Complex Organizations*, Addison Wesley, Reading, MA.

Håkanson, Håkan (1989) *Industrial Technological Development*, Routledge, London.

Hannan, Michael T. and Freeman, John (1977) The population ecology of organizations, *American Journal of Sociology*, No. 72.

Hax, Arnoldo C. and Majluf, Nicolas S. (1991) *The Strategy Concept and Process*, Prentice Hall, Englewood Cliffs, NJ.

Hickson, David J., Hinnings, C. R., Lee, C. A., Schenck, P. E. and Pennings, J. M. (1971) A strategic contingencies theory of intraorganizational power, *Administrative Science Quarterly*, Vol. 16, June, pp. 216–229.

Ioannides, Stavros (1992) *The Market, Competition and Democracy*, Edward Elgar, Aldershot.

Laffont, Jean-Jacques and Tirole, Jean (1993) *A Theory of Incentives in Procurement and Regulation*, MIT Press, Cambridge, MA.

Lane, Jan Erik (1993) *The Public Sector, Concepts, Models and Approaches*, Sage, London.

Leavitt, Harold J., Dill, William R. and Eyring, Henry B. (1973) *The Organizational World*, Harcourt Brace Jovanovich, New York, NY.

March, James G. and Olsen, Johan P. (1976) *Ambiguity and Choice in Organizations*, The Norwegian University Press, Bergen.

Mayo, Elton (1945) *The Social Problems of an Industrial Civilization*, Graduate School of Business Administration, Harvard University, Boston, MA.

Merton, R. K. (1957) *Social Theory and Social Structure*, Free Press, Glencoe, IL.

Meyer, John, W. and Rowan, Brian (1977) Institutionalized organizations: formal structures as muth and ceremony, *American Journal of Sociology*, Vol. 83, pp. 340–363.

Musgrave, R. A. (1959) *The Theory of Public Finance*, McGraw-Hill, New York, NY.

Musgrave, R. A. and Peacock A. T. (eds) (1967) *Classics in the Theory of Public Finance*, St. Martin's Press, New York, NY.

Niskanen, W. A. (1971) *Buraucracy and Representative Government*, Aldine-Atherton, Chicago, IL.

Norman, Michael and Stoker, Barry (1991) *Data Envelopment Analysis: the Assessment of Performance*, Wiley, Chichester, NY.

Orru, Marco (1994) *The faces of capitalism*, paper presented at the Sixth Annual International Conference on Socio-Economics, HEC, Paris.

Parkinson, N. C. (1957) *Parkinson's Law or the Pursuit of Progress*, John Murrey, London.

Perrow, Charles (1970) *Organizational Analysis: A Sociological View*, Wadsworth, Belmont, CA.

Perrow, Charles (1974) Is business really changing?, *Organizational Dynamics*, Summer, pp. 31–44.

Pfeffer, J. and Salancik, G. R. (1978) *The External Control of Organisations*, Harper and Row, New York, N.Y.

Piore, Michael J. and Sabel, Charles F. (1984) *The second industrial divide: possibilities for prosperity*, New York, Basic Books.

Porter, Michael (1990) *The Competitive Advantage of Nations*, Macmillan, London.

Powell, Walter W. (1981) Neither market nor hierarchy: network form of organization *Research in Organizational Behaviour*, Vol. 12, pp. 295–336.

Pugh, D. S., Hickson, D. J. and Hinnings, C. R. (1969) An empirical taxonomy of structures of work organizations, *Administrative Science Quarterly*, Vol. 14, March, pp. 115–126.

Roethlisberger, F. J. and Dickson, William J. (1939) *Management and the Worker*, Harvard University Press, Cambridge, MA.

Samuelson, P. A. (1954) The pure theory of public expenditure. *Review of Economics and Statistics*, Vol. 40, pp. 387–389.

Schumpeter, Joseph (1943) *Capitalism, Socialism and Democracy*, George Allen & Unwin Ltd., London.

Scott, Richard (1981) *Organizations: Rational, Natural and Open Systems*, Prentice Hall, Englewood Cliffs, NJ.

Selznick, Philip (1949) *TVA and the Grass Roots*, University of California Press, Berkeley, CA.

Starbuck, W. H. (1965) Organizational growth and development. In March, J. G. (ed.) *Handbook of Organizations*, Rand McNally, Chicago, IL.

Stinchcombe, Arthur L. (1984) *Contracts as Hierarchical Documents*, Work Report No. 65, Institute of Industrial Economics, Bergen.

Swinth, Robert L. (1974) *Organizational Systems for Management: Designing, Planning and Implementation*, Grid, Columbus, OH.

Taylor, Fredrick W. (1911) *Principles of Scientific Management*, Harper, New York.

Thompson, James D. (1967) *Organizations in Action*, McGraw-Hill, New York, NY.

Trist, E. L., Higgin, G. W., Murray, H. and Pollock, A. B. (1963) *Organizational Choice*, Tavistock, London.

Udy, Stanley H., Jr. (1959) Bureaucracy and rationality in Weber's organization theory, *American Sociological Review*, Vol. 24, December, pp. 791-95.

Weber, Max (1964) *Social and Economic Organization*, Collier Macmillan, London.

Weick, Karl (1969) *The Social Psychology of Organising*, Addison-Wesley, Reading, MA.

Whitley, Richard (1992) *European Business Systems: Firms and Markets in their National Contexts*, Sage, London.

Williamson, Oliver (1975) *Markets and Hierarchies: Analysis and Antitrust Implications*, Free Press, New York, NY.

Woodward, Joan (1965) *Industrial Organization: Theory and Practice*, Oxford University Press, New York, NY.

Part Two

Varieties of Liberal Reform

Chapter III
The British Market Reform:
a Centralistic Capitalist Approach

STEVE THOMAS

I. Introduction

Of all the reforms to electricity supply industries worldwide, that which has been undertaken in England and Wales[1] has attracted most attention. It was not only the first to undergo major restructuring, but in some respects it remains the most radical in scope. It involved revolutionary changes in the way the system operated, including the introduction of competition as the driving force for much of the industry, a transfer of ownership from central government to the private sector, and the break-up of the company at the centre of the old system, the Central Electricity Generating Board (CEGB). The British experience therefore provides an essential reference point for any electricity supply industry reform. Figs III.1 and III.2 show the structure of the electricity supply industry before and after privatisation.

II. The Pre-Reform System[2]

Like most electricity supply industries in Europe, the UK industry evolved as a mixture of public and private ownership franchised by

---

[1] While the title of this chapter refers to Britain, in fact only the changes relating to the system for England and Wales will be covered in detail. There have been changes of structure in the systems that supply Scotland and Northern Ireland but scope for reform was seen as more limited because of the relatively small size of these systems. The changes have therefore been far less radical than those applied the system for England and Wales. For example the changes for Scotland have largely been a transfer of ownership from the public to the private sector for all except the nuclear plants with little attempt to introduce competition.

[2] For a comprehensive account of the history of the British electricity supply industry up to 1979, see Hannah (1979) and Hannah (1982). See also The Electricity Council (1987).

Generation	**CEGB, Imports**
Dispatching	**CEGB**
Transmission	**CEGB**
Distribution and Supply	**12 Area Boards**

Fig. III.1. The electricity supply industry in England and Wales before privatisation. *Note:* Very large consumers were able to negotiate direct supply contracts with the CEGB.

Generation Imports	**National Power, PowerGen, Nuclear Electric, IPPs,**
Dispatching	**The Pool**
Transmission	**National Grid Co.**
Distribution	**12 Regional Electricity Companies**
Franchise Supply	**12 Regional Electricity Companies**
Non-Franchise Supply	**Licensed generators, 12 RECs, other licensed suppliers**

Fig. III.2. The electricity supply industry in England and Wales after privatisation. *Note:* From Vesting Day (1 April, 1990) all consumers with a maximum demand over 1 MW were able to negotiate the supply of power from any licensed supplier. From 1 April, 1994, this limit was reduced to 100 kW, and from 1 April, 1998, it is planned that all customers will be able to choose their supplier.

local (municipal) authorities. Even before nationalisation in 1947, the industry had moved towards a greater degree of public ownership. The institution which controlled the dispatch of plant since a national grid was first established in 1933, the Central Electricity Board, was a public corporation, and much of the supervisory and planning functions were carried out by public bodies.

When the Labour Government nationalised the industry in 1947, it was apparent that Government had to take decisive steps to overcome a chronic shortage of generating capacity caused by World War II, if post-war economic recovery was not to be hindered. Opposition to nationalisation was therefore muted and, from then up to the period of the

Thatcher Governments, state ownership of the electricity supply industry was not a party-political issue.

The basic structure established in 1947, of a number of distribution companies (known as area boards) and one generation and transmission company (then the Central Electricity Authority (CEA)) survived through to privatisation. In 1955, the southern Scotland system was separated off,[3] leaving 12 area boards. In 1957, the generation and transmission company covering England and Wales was renamed the Central Electricity Generating Board (CEGB). The Electricity Council, a central co-ordinating and marketing body representing the CEGB and the area boards, was also established then, partly to provide central services for the companies but also to lessen the dominating influence of the CEGB. Despite this, after nationalisation, the CEA and then the CEGB were indisputably the key body for the whole of the UK in making all major investment, technology and fuel-choice decisions.

II.A. The record of the CEGB

The UK national grid was one of the first to be established, and the CEGB's operation and maintenance of it has generally been regarded as highly effective. However, in other respects, the record of the CEGB is more questionable.

II.A.a. Purchasing

Like other nationalised utilities, the CEGB rapidly became a large, powerful organisation dominating most of its suppliers. The nationalised UK coal-mining industry, the National Coal Board (NCB), later the British Coal Corporation (BCC), became increasingly dependent on the CEGB as its other markets for coal declined. It was also normal practice amongst those countries with a power-station equipment-supply industry that all major items of equipment would be purchased from home sources, and the CEGB followed this practice. The result of this was that suppliers of such equipment were generally dependent for 80% or more of their sales on their home market (Thomas and McGowan, 1990). This gave the CEGB immense (effectively monopsony) power over its equipment suppliers, with its purchasing policy determining which technologies were developed and which companies survived. The CEGB built up design expertise in much of the equipment it purchased, and contributed significantly to the design of the whole range of its equip-

---

[3] The Scottish system comprised two vertically integrated regional companies responsible for generation, transmission and distribution.

ment purchases from turbine generators to rail wagons for coal transportation.

However, while in theory it had power over its suppliers, it was often unable or unwilling to exercise it fully. It was effectively banned from importing coal and could not substitute oil for coal without government approval. It therefore had little scope to exert downward pressure on the prices it paid for British coal. Equally, to choose not to buy equipment from one manufacturer would, in many cases, effectively have put that supplier out of business. This was not a comfortable position for a nationalised industry to be in, and major equipment purchases were often distributed amongst the suppliers on a so-called 'Buggins turn' basis (i.e. distributed evenly amongst them) on largely 'cost plus' terms. Effectively, therefore, the CEGB was able to exert heavy technological influence on its suppliers, but little economic pressure.

A 1981 investigation of the CEGB by the UK Monopolies and Mergers Commission appeared to be sceptical about how healthy the relationship between the CEGB and its suppliers was, noting a lack of confidence between the CEGB and NCB and expressing doubts as to whether it was appropriate for electricity consumers to have to pay the premium which UK equipment appeared to command over world prices (Monopolies and Mergers Commission, 1981, pp. 290–291).

II.A.b. Site management, planning and investment
The post-war plant shortage had been overcome by the mid-1950s, but by the early 1960s, under-forecasting of demand had resulted in the re-emergence of power shortages, with many power-cuts in the very cold winter of 1962/63. Failure to keep the lights on was then probably the worst humiliation for an electricity supply company, and the credibility of the CEGB with the public and government was severely damaged. Largely as a result, for about the next 15 years, forecasts of demand erred significantly on the high side, with the industry consistently failing to respond to signals that the days of annual electricity demand growth of 7% were over. From 1961 to 1980, the CEGB's demand forecasts for five to six years forwards (its planning period) were far too high. While many utilities were caught out by the fall in energy demand which the first oil crisis of 1973/74 caused, CEGB overforecasting was actually at its most inaccurate in the mid-1960s, when demand was over-estimated by more than 35% (Select Committee on Energy, 1981, pp. 19–20).

While this did lead to some over-capacity in generation, ironically the CEGB's increasing problems in controlling construction times meant that at least part of the forecasting error was masked. Power station construction began to go conspicuously wrong in the mid-1960s, when the first of the power stations using 500 MW steam-turbine generators began to enter service. These frequently exceeded their lead-time schedules because of poor management of on-site construction and led to a major government investigation, the Wilson Enquiry, into what was going wrong.[4] The 500 MW units also proved extremely unreliable in their early years. This unreliability necessitated several increases in the target reserve margin of plants because the availability of these new units at the time of peak demand was so low. This extra margin also served to absorb some of the plant which would otherwise have been seen as surplus.

Far from improving after the Wilson Enquiry, control of industrial relations at construction sites deteriorated to the extent that, in 1976, at the Grain site, where a five-unit oil-fired station was being built, construction work was suspended after a series of disputes. The station was finally completed in 1979–1983, 5–9 years later than scheduled (Monopolies and Mergers Commission, 1981, pp. 265–267). Other sites of conventional power-plant construction fared little better.

The situation on the nuclear construction sites was even worse, compounded by a poor choice of technology (Williams, 1980). In 1965, the UK Government chose to adopt the indigenous advanced gas-cooled reactor (AGR) design as the successor to the first generation of nuclear power plants, the Magnox design. The implementation of the AGR programme was described in 1973 by the then Chairman of the CEGB as 'a catastrophe we must not repeat' (Select Committee on Science and Technology, 1974). As well as poor management of on-site construction, the problems with the AGRs included the intrinsic weaknesses of the technology and inadequacies in the procurement strategy. While some of the blame for these errors lies with the Government (which made the decision on technology choice and on procurement strategy) and the UK Atomic Energy Authority (which was a powerful advocate for the AGR), the CEGB, as manager of construction and procurement, cannot escape all responsibility (Rush *et al.*, 1977; Henderson, 1977).

By the early 1980s, the over-forecasting of demand in the 1970s meant that little new capacity had been ordered for some time, and there were few remaining sites on which power-station construction was taking place. Nevertheless, on these few sites, there was some evidence that schedules were being more closely adhered to and that the CEGB was beginning to solve its site-management problems.

Even as late as 1980, the UK Monopolies and Mergers Commission

---

[4] See Report of the Committee of Enquiry into Delays in the Commissioning of CEGB Power Stations (1969) and National Economic Development Council (1970).

(MMC) in its report made serious criticisms of the CEGB's planning and investment appraisals. This investigation uncovered serious 'appraisal optimism' in its treatment of nuclear power. Effectively, the MMC found that the CEGB's planning procedures were systematically biased in favour of nuclear power and the MMC found it necessary to condemn the CEGB's procedures in the strongest terms available to it, that its investment appraisals were 'seriously defective and liable to mislead' and therefore operated 'against the public interest' (Monopolies and Mergers Commission, 1981, pp. 292–293).

Chastened by this catalogue of failures and the mounting public criticism of its operation, the CEGB was taking clear remedial steps. It had begun to take a more aggressive stance against its suppliers of fuel and equipment, and labour relations at its construction sites were less problematic. Forecasting accuracy improved, and the case it put forward for the Sizewell B PWR in 1982 was based on much more realistic demand forecasts and cost projections than earlier decisions. In addition, labour productivity, fuel efficiency and other key performance indicators continued to show steady progress from 1970 onwards.

Nevertheless, new criticisms arose, particularly that the CEGB was too monolithic and protected its effective monopoly too vigorously. It still appeared single-minded in its pursuit of nuclear power almost to the exclusion of all other options, and a 1983 Government initiative to promote industrial cogeneration received little encouragement from the CEGB (Vickers and Yarrow, 1988, p. 292).

II.A.c Overall assessment

Overall, the CEGB was a far from perfect organisation, but it could certainly not be said to have failed and was probably no worse an organisation than most of its international peers. At the time of privatisation, a number of commentators who had been critical of the performance of the CEGB argued that the problems of the organisation could have been overcome without the upheaval of a major restructuring. The CEGB itself argued strongly against privatisation, citing the risk to security of supply if it was restructured, in particular if it lost control of the high-voltage transmission grid. However, past mistakes did damage its reputation and meant that opposition to its break-up was weaker than in previous utility privatisation, a factor which surprised the CEGB, which expected to be able to count on a surge of public support which would prevent its break-up.

Another factor influencing government towards privatisation was that increasingly, government decisions on power supply carried little kudos and often severe unpopularity. In the early post-war years, taking control of the provision of vital utility services and launching technological initiatives such as nuclear power generally gave prestige to the government. Now, government-inspired decisions on technology, siting and prices are all far more likely to lose than to gain votes. In this light, privatisation can be seen as allowing the market to do what was, till then, the government's 'dirty work'. For example, electricity privatisation allowed decisions which inevitably led to the run-down of the British coal industry but which could be passed off as the choice of the market. However, other political forces provided the main impetus for restructuring, in particular the Thatcher Governments' policy of transferring state-owned assets to the private sector.

III. The Privatisation Programme[5]

While, from outside, the impetus for privatisation might appear to have been driven by a belief that a privately owned, fully competitive industry structure would be more efficient than a monopoly state-owned industry, the reality was much more complex. Indeed, for utilities the element of creating competitive structures came quite late to the process.

In the first five years of the Thatcher Government, from 1979/80 to 1983/84, £2.2 billion was raised by sales of shares in state-owned assets, but all involved companies were already operating in competitive markets. Of this, 37% was accounted for by sales of government-owned shares in British Petroleum, 28% by the privatisation of Britoil, an oil exploration and production company, and 20% by the privatisation of Cable and Wireless, a telecommunications services company. State ownership of these companies had been justified on the grounds of strategic macro-economic objectives, for example of retaining and developing skills or controlling the depletion of a national resource. It was assumed that no restructuring and no new regulatory mechanisms would be required by these asset disposals because the companies operated in competitive markets.

One of the main motives behind these sales seems to have been a belief that a simple transfer of ownership to the private sector would improve the efficiency of the company. Another motive was to generate cash for the Treasury and to remove their capital spending from the public sector borrowing requirement. In the early 1980s, the UK economy was in deep recession and the proceeds of sales of state assets were used to mitigate its effects on public-sector finances. Even when the UK had emerged from recession in the mid-1980s, the government was committed to cuts in direct taxes, and privatisation was able to

---

[5] Kay *et al.* (1986) and Vickers and Yarrow (1988).

help finance this. At the time of electricity privatisation, the Treasury had a target of proceeds of £5 billion a year from sales of assets.

It was only with the privatisation of British Telecom in November 1984 and British Gas in December 1986 that the Government started to take on the more difficult task of privatising utility industries. This second phase of privatisation, sale of utility industries, could have raised emotive issues of whether vital public utility services such as water and power could be safely left to the market. But the gas and telecommunications industries were sold intact, with their structures and their monopoly position largely unaltered. This gave the impression to the public, at least at the time, that little of consequence had changed. Regulatory bodies were created (OFTEL and OFGAS, respectively) but these excited little public interest.

Leaving the monopolies intact gave these companies a very secure future and meant that revenues from privatisation were much higher than if aggressive competition had been expected. It also meant that shares could be sold to the general public with little of the risk which the ownership of shares usually entails. Indeed, instant profits were available to those buying shares of these companies as the price of shares rose sharply after flotation. The two-stage sale of British Telecom raised £3.9 billion, and the sale of British Gas raised £5.6 billion. However, there was increasing scepticism about the 'magic' that simply transferring assets from state to private ownership could work. There was public tension between the regulators and their industries as the regulators tried to impose competitive forces on the utilities. By the time electricity privatisation was proposed in 1987, it was clear that this time a simple transfer of assets from public to private ownership would not be acceptable to the free-market wing of the Conservative Party.

The thinking behind the model which emerged for electricity now provides the basis for utility liberalisation, including the gas, rail and telecommunications sectors for which a natural monopoly supply network is a major part of the industry. Since the completion of the privatisation of electricity, privatisation of the remaining utility industries, including water (privatised in 1989), rail (expected to be privatised in 1996) and the Post Office (on which privatisation proposals were being formulated but were abandoned after strong opposition) has been addressed. British Coal has also been privatised. A third phase of privatisation, the 'market testing' of services for public sector responsibilities, ranging from catering in hospitals to owning and operating prisons, now provides the main focus for the privatisation programme.

III.A. Electricity privatisation

The size and complexity of the electricity supply industry meant that there could never have been any doubt that it would be the most difficult privatisation attempted. The asset value of the UK electricity supply industry was estimated to be approximately 'four times as large as the total asset base of all the industries which were privatised in the first two Thatcher terms' (Holmes *et al.*, 1987). Instead of dealing with just one company, as was the case with telecommunications and gas, the electricity supply industry for England and Wales was made up of 14 companies, 12 regionally based distribution companies, the CEGB and the Electricity Council. In addition, with electricity demand-growth limited (see Table III.1), the new companies would only be able to grow by diversifying or competing successfully with each other.

While the government was clear about what it did not want—a simple transfer of assets from public to private ownership with no new competitive mechanisms—it was far from clear about what it did want. Despite contemplating momentous changes to one of the most fundamental sectors of the economy, there was no vision of how the final structure would look, nor has there been a constant guiding hand providing continuity. The Conservative Party manifesto of 1987 contained a commitment to privatise the electricity supply industry, but few details on how it would be done. It also contained a commitment to promote nuclear power development, a commitment which many predicted would be difficult to reconcile with privatisation (Chesshire, 1989).

A clear additional goal behind privatisation was to break once and for all the power of the coal-mining unions. Whilst this was not admitted at the time, subsequent memoirs from the Cabinet Ministers involved

Table III.1. Electricity demand in the UK by sector (TWh)

	1970	1975	1980	1985	1987	1989	1991	1993	1994
Fuel Industries	7	6	7	8	8	7	7	7	5
Industrial	73	75	80	80	83	90	90	88	90
Transport	3	3	4	3	3	3	5	7	7
Domestic	77	89	86	88	93	93	98	100	101
Public administration	12	13	16	17	18	19	20	21	22
Agriculture	4	4	4	4	4	4	4	4	4
Commercial	24	28	36	42	47	53	54	56	56
Total	199	219	232	242	256	267	278	283	284

Source: Digest of United Kingdom Energy Statistics (various years).
Note: A full series of separate demand data is not available for England and Wales.

have confirmed this motive.[6] The protected market for British coal could not be expected to be sustained in a private competitive system, and private companies would not be so vulnerable to the political pressures which weakened the electricity supply industry's negotiating position against the British coal industry.

The election manifesto commitment effectively meant that the electricity supply industry had to be sold. The option that, after careful consideration, retaining the industry in public ownership was the best choice was not one that was available to the Secretaries of State for Energy charged by Mrs Thatcher with carrying through the process. Decisions which could be represented as policy 'U-turns' had become anathema to the Conservative Government, and the Treasury could not be denied its pound of flesh. In addition, the process of privatisation would have to be completed within the time span of one Parliament. Any possibility that the process would have remained uncompleted when a General Election was held would have been far too risky, both politically and commercially. The revenues accruing could also be used to party-political advantage by cutting taxes just prior to a General Election. While a British Parliament can run for up to five years, in general the period between elections (which is largely at the discretion of the Government) is usually expected to be no more than four years. Allowing for the period leading up to an election when little of substance can be accomplished, and that the structure had not been chosen in 1987, this meant that the whole process, from design of the new structure to sale of the companies, had to be crammed into a period of less than four years.

The government's February 1988 White Paper on privatisation finally gave some details of the structure proposed (Department of Energy, 1988; Scottish Office, 1988). The distribution companies, renamed Regional Electricity Companies (RECs), would be sold intact. However, they would be required formally to separate the distribution business, that is the operation of the local physical infrastructure (the 'wires business') from the supply business, that is the purchase of electricity, marketing, meter-reading and consumer billing. The wires business was seen as a natural monopoly, while the supply business could be opened to competition.

The CEGB would be divided into three parts. The generation side would be split into two competing companies, one containing 70% of

---

[6] Several former Government ministers in Thatcher Governments have stressed the deep animosity between the NUM and the Government and the Government's strategic objective of breaking NUM power. See for example, Ridley (1991), Walker (1991), Parkinson (1992) and Lawson (1992).

capacity, including the nuclear plant, and the other containing 30% of capacity. The transmission business, seen as a natural monopoly, was to be operated as a third separate company.

As with previous privatisation of utility industries, regulation, where necessary, would be under the aegis of a Director General, in this case, of Electricity Supply (the DGES), with the assistance of a specific new regulatory body, the Office of Electricity Regulation (OFFER). It was assumed that much of the industry, including generation and ultimately supply, would be sufficiently competitive not to require formal regulatory procedures, but where formal price regulation was required, in the distribution and transmission sectors and initially in supply, it would be based on incentive regulation using the simple 'RPI−x' formula. 'RPI', or retail price index, is a measure of consumer price inflation, and 'x' is an incentive term. Put simply, this means that the price of the elements under the control of the utility are allowed to rise by the general rate of inflation minus a small term which the utility must cover by improving its efficiency. Any changes in the cost of the elements of the total price which are seen to be outside the control of the utility can be passed on in full to consumers.

The basic philosophy underlying the new structure was that all areas of electricity supply which are not natural monopolies would be opened to full and vigorous competition as soon as was practical. The remaining area, which amounts to the physical infrastructure of the network, the high-voltage transmission system and the local distribution system, would be regulated by the DGES, using incentive regulation.

The most obvious area which could be opened to competitive forces was generation. The Power Pool was to be the main price-setting arena—contracts of limited term outside the Pool were anticipated for power purchasers who needed greater predictability in their costs, but pricing of these contracts would tend to use Pool Prices as their bench mark. These mechanisms were expected to be competitive enough that no need for routine regulation of the generation sector was anticipated. Market forces were also expected to remove the need for planning generation capacity.

It was also anticipated that elements of the supply of power to final consumers could be made fully competitive. Clearly there was no question of duplicating the physical distribution infrastructure, but if the network was a resource open to all, on non-discriminatory terms, competition could be introduced. For example, if a supply company could purchase power more economically, or it could supply customers incurring lower overheads, it would be able to offer cheaper terms to consumers. From Vesting Day, April 1, 1990, (the day the new structure came into being), consumers with demand greater than 1 MW were able to negotiate their supply from any licensed supplier. This limit was reduced to 100 kW in April 1994, and is scheduled to be removed altogether in April 1998. Customers negotiating the supply of power on individual contracts are assumed to have open to them a sufficiently competitive market that regulation is not required. However, as a back-up for those not wishing to exercise choice, the local REC is obliged to offer supply to all consumers with demand less than 10 MW under the terms of a published and regulated tariff.

Details of how the system would run in practice were still sketchy at this point, but a number of factors conditioned the structure and how it was ultimately made operational. Some of these have been discussed earlier, including the need for the new structure to be competitive and the necessity to complete privatisation within one Parliamentary term. Other factors are detailed below.

III.A.a. A hidden agenda on coal and nuclear[7]

The Conservative Party's manifesto commitment to nuclear power was noted earlier. This stems in part from the fact that, as with many right-of-centre parties worldwide, the Conservative Party has always tended to be well disposed towards nuclear power. In addition, the defeat of the coal-miners' strike in 1985 was in no small part due to the contribution of nuclear power,[8] and the Conservative Party was seen to owe nuclear power a debt. Privatisation was not expected to help the nuclear industry by itself—the CEGB was already fully committed to nuclear power—but protection for the nuclear industry had to be built into the new structure. Indeed, thinking on the generation side was based almost entirely on finding a structure which would allow nuclear power to flourish.

By contrast, there was a long history of animosity between the Conservative Party and the main coal-miners' union, the National Union of Mineworkers (NUM), particularly following two strikes in the early 1970s and other disputes which inflicted serious political damage on the Conservative Party. The Conservative Party was unhappy about the high dependency of the electricity supply industry on coal (about 80% of power was generated using British coal) which seemed to give miners scope to thwart Government wishes by the use of industrial action.

---

[7] In Parker and Surrey (1992), the authors chart systematic Government discrimination against British coal and in favour of nuclear power in the period of the Thatcher Governments.

[8] For example, many of the nuclear power plants were operated at up to 10% above their nominal capacity during the winter of 1984/85. See International Atomic Energy Agency (1986).

While the wish to break the power of the NUM could not be stated in those terms, the Government used code-words such as 'diversification' which few failed to decode (Lawson, 1992, p. 168).

III.A.b. Poor understanding of the sector by the government and those advising it

The Department of Energy in the UK was not regarded as one of the more effective government departments, and it is significant that since it was abolished following the General Election in 1992, there has scarcely been a voice raised lamenting the loss. The most crushing indictment of its lack of understanding of the sector was its failure to alert the government to the total impracticality of the plans to privatise nuclear power. The private sector, with little experience of dealing with nuclear power, was no more impressive, and investment analysts did little to alert the government to the problems in its approach.

III.A.c. Regulation had to be light

In part, this was a reaction against the US system of economic regulation, which was perceived to be cumbersome, bureaucratic and expensive and was too dominant a force in utility decision-making. There was also perhaps a genuine belief that a market could be created which was competitive enough not to require heavy regulation. More practically, from the point of view of public presentation, it would have been difficult to argue that the private sector had major advantages over public ownership if a heavy regulatory structure was required to ensure that the private sector did not abuse its market power.

III.B. The period of electricity privatisation

The period of privatisation was marked by false starts, leaks of plans and failure to meet target dates. In part, this reflected the naiveté of many of those involved in privatisation. More charitably, it was a reflection of the fact that what was being attempted was immensely complex and had no precedent. The UK was the first country in the world to try to design an electricity supply system in which competition was the central feature.

The most conspicuous and embarrassing failure during the process of privatisation concerned nuclear power. The proposed structure for the generation side of the business envisaged for England and Wales was tailored so that the nuclear power plant could be accommodated in the private sector. It involved the creation of only two generating companies, one of which was to be large enough to have the economic and technical strength to deal with the nuclear power plant, which then comprised about 10% of capacity. It was originally planned that this company (which became National Power) would take 70% of the CEGB's generating capacity. If this company was not to be so large as to totally dominate the market, it was argued that the other 30% had to be concentrated in only one other company (PowerGen) to provide effective competition to the bigger company. The non-nuclear power stations were shared out very carefully on the basis of the fuels used and the location, so that neither of the two generators had a significant advantage in its cost base.

However, the Magnox plant had very little operating life left and massive liabilities for decommissioning and spent-fuel disposal, and the AGRs then showed little sign of becoming reliable generators of power. For the future, the higher rate of return required by private markets (now estimated to be more than 12% real *per annum*) compared to that required from state-owned companies by government (during privatisation raised from 5% to 8%) seemed likely to make new nuclear power stations uncompetitive. Far from being the cheapest generation option as the CEGB had long argued, using private-sector investment-appraisal parameters, it turned out that power from the Sizewell B PWR would be approximately twice as expensive as the alternatives. These factors were obvious to almost all experienced energy analysts. In April 1987, before the Conservative Party was re-elected, Holmes *et al.* wrote: 'It seems the only possible route is to leave the nuclear industry in state hands' (Holmes *et al.*, 1987, p. 64). In July 1988, the House of Commons Select Committee wrote: 'The independent witnesses we examined were unanimous in their view that ... private generating companies would be most unlikely to build new nuclear power plants' (Select Committee on Energy, 1988).

There was some acknowledgement that the high capital costs and the commercial risks of nuclear power might mean that generating companies would not opt for nuclear power. A 'non-fossil fuel obligation' (NFFO), expected to be about 20%, was therefore to be placed on distribution companies which would require them to obtain a certain percentage, to be specified by the relevant minister, of their power from non-fossil fuel sources. If distribution companies could see they would not be able to meet this target in the future, perhaps due to plant retirement or demand growth, they would be obliged to contract someone to build additional non-fossil fuel capacity. The possibility of a subsidy for nuclear power, a 'fossil fuel levy' (FFL), was also then raised, but its size and method of operation were not specified.

It soon became clear that long-term obligations which constrained their commercial decision-making could not be placed on private-sector companies, and the NFFO quickly became no more than a mechanism to ensure that existing nuclear power plant was used to its full extent. This change in policy was particularly embarrassing to the CEGB, which at that time was arguing the case for a new PWR at a public inquiry on the basis that the plant would be needed to fulfil the expected level of the NFFO.[9]

Why it took two years for the government to act on these and many other similar warnings that nuclear power could not be sold is not clear. In part, it must be due to the naiveté of the Department of Energy, but it may also have been due to the unwillingness of senior politicians to listen. The warnings may have been interpreted as ideologically motivated statements by parties with an interest in obstructing the process of privatisation. Even when the facts could be ignored no longer, the Government retreated slowly. In July 1989, the then Secretary of State for Energy and the architect of the privatisation proposals, Cecil (now Lord) Parkinson, in practically his last act in that role, withdrew the Magnoxes from the sale.

By then, delays to electricity privatisation seemed so serious that it might have to be abandoned until after the next election. Parkinson was replaced by John (now Lord) Wakeham. This move was seen by much of the media as the replacement of a weak minister by a 'political fixer'. The emphasis switched from trying to create the most competitive market solution to simply taking whatever steps were necessary to complete the job in the time available. The remainder of the nuclear power sector was withdrawn from the sale in November 1989. However, the demands of the timetable were such that no more than a minimum of re-thinking was possible. No structural changes were introduced, except that National Power was to be privatised without the nuclear power plant. This was now to be held by a new, publicly owned company, Nuclear Electric. Wakeham has since claimed that had it been known from the outset that nuclear power could not be privatised in the sort of competitive structure proposed, it would have been preferable to split up the generation business into much smaller units.[10]

---

[9] The initial CEGB case to build Hinkley Point C stated 'the policy for diversity justifies consent for and the construction of generating plant which would be needed to meet the requirement for non-fossil-fuelled capacity. It is the CEGB's case that Hinkley Point "C" is essential to meet that requirement.' Central Electricity Generating Board (1988) p. 10, para. 1.13. The CEGB sought to avoid comparing the cost of power from a PWR with that of a coal-fired station, but the inquiry inspector concluded that 'using a 10% discount rate [the rate he concluded was appropriate for the private sector], there would be a significant advantage in favour of coal fired plant' (Barnes, 1990, Vol. 3, p. 856, para. 29.138).

[10] In evidence to the Energy Select Committee, Wakeham said 'If I was starting from scratch I would not have decided to split the CEGB fossil stations into two companies' (Select Committee on Energy, 1990, p. x, para. 6.

Nuclear Electric was to be compensated for the high cost of nuclear power from the proceeds of a fossil fuel levy (FFL) of about 10% charged on all sales of power. The FFL has yielded between £1.2 and £1.4 billion *per annum*, and has provided 40–50% of Nuclear Electric's income. Privileges under the NFFO and the Existing Nuclear Operating Regime (ENOR) effectively meant that nuclear power plants would be used whenever available. However, a moratorium was placed on all new orders for nuclear power plant, excluding the Sizewell B PWR which was then already under construction, until at least 1994, when a Government review of nuclear power policy was to be completed. The European Commission forced one major concession on these proposals, that the FFL had to be phased out by 1998.

A particular concern was that stability should be maintained to power supplies and prices in the settling-down period after the end of the old system and the establishment of the new system. This was to be ensured by the Energy Minister imposing three-year contracts between the new generating companies and the RECs for the supply of power at pre-set prices and between the generating companies and British Coal for most of the coal that was expected to be needed.

For the RECs, this meant that the price of most of their power needs was fixed for the first three years, independently of what happened in the Pool. The generators still had to bid to get their plant used, but if the Pool price was below the contract price, under the terms of these so-called contracts for differences, the RECs reimbursed them the difference and vice versa. The Pool was inevitably therefore of little relevance to the price of power for at least the first three years of its operation. The main area of competitive activity was in the market for direct supply to customers whose maximum demand exceeded 1 MW. Such customers were free to negotiate their power purchases with any REC or generator company. Given the similarity of the cost base of the two generators, and that the RECs were buying their power from these generators at cost-related prices, there was clearly little scope for the prices offered to these large consumers to vary by much. Nevertheless, the new companies were keen to be seen to be competing successfully in the new market, and there was a very public process of competition as the new companies vied with each other to win the contracts.

The first three-year coal contracts committed the generators to continue to purchase about the same volume of coal as in previous years from British Coal. However, the prices were to fall significantly in real terms, so that at the end of the contracts, British Coal would be supplying coal at around world prices. This three-year period was intended to give British Coal time in which to adjust to the new commercial environment, and also to prepare British Coal itself for privatisation.

Another area causing problems in the period up to privatisation was the Power Pool. This was to be the centre-piece of the system which would be an open, transparent spot-market for power which would mean that all generators, large and small, would be able to compete on equal terms to sell their power. Buyers would also be able to place bids specifying the price at which they were prepared to purchase power. In autumn 1989, under the more pragmatic direction of Wakeham, the proposed new 'double' Pool structure had to be abandoned because of the complexity of the computer software required. It was replaced by a new system which necessarily was much simpler so that it could be in place by April 1, 1990, Vesting Day, when the new companies came into formal existence and the Power Pool began operation. The feature of demand-side bidding was lost, and the Pool became a one-sided market based on the software the CEGB had used to despatch its plants.

In December 1990, the RECs were sold to the public in a share issue which was heavily over-subscribed. The share price of the new companies rose rapidly, effectively generating instant profits for those purchasing shares. The transmission sector was hived off into a new company, the National Grid Company (NGC).[11] Ownership of NGC was passed originally to the 12 RECs, but in December 1995, the DGES required the RECs to sell most of their shares in the company *via* a stock-market flotation. The risk that ownership by the RECs of the company controlling the grid, particularly the despatching system, would lead to grid policies which would unfairly favour the RECs was countered by provisions which restricted the extent to which the RECs could influence the policies of the NGC.

In March 1991, 60% of the shares in the two generating companies were sold, again in a heavily over-subscribed sale, and at prices well below those that the shares quickly rose to. The other 40% of shares remained in UK Government hands until they were sold early in 1995.

IV. The New System in Practice

IV.A. The pool

If the UK experiment was to lead to a competitive electricity supply system, a genuinely competitive Pool was essential. A two-sided market including buyers as well as sellers proved impossible to implement in the time available, but even the less ambitious generation Pool had

---

[11] Most of the grid operates at 275 kV or 400 kV. The distribution system, owned by the RECs, operates at 132 kV or less.

the potential to have a major impact on the competitiveness of the system.

The Pool was to be a spot-market for power covering all generating units above a minimum size. Only units which have placed successful bids with the Pool can be operated. Bids must be placed by 10 a.m., specifying a price at which the owners of the unit are prepared to supply power for the next day. The basic principle of the Pool is that the lowest bids needed to meet the demand forecast on a half-hourly basis are chosen, producing a different Pool price for each half hour of the next day. The highest successful bid (the system marginal price) is the main component of the Pool purchase price which all successful bidders receive. Forecasts of the future system marginal price were expected to be the main mechanism for attracting new generators: a high system marginal price would provide a signal to entrepreneurs that profits could be made from generation, and they would be encouraged to build new capacity or make investments in replacing expensive capacity with cheaper plant. A low system marginal price would 'shake out' inefficient capacity.

The major addition made to the system marginal price in order to arrive at the Pool purchase price is a capacity payment. This is intended to compensate and give incentives to owners of peaking plant, and the size of this payment is therefore designed only to become significant when the volume of plant bid is close to that required. When the gap between supply and demand is small, this capacity payment rises very steeply and, at the point at which supply is no longer sufficient to meet demand, reaches more than £2/kWh. By this mechanism, owners of peaking plant, such as open-cycle gas turbines, which would only be expected to operate for a few hours a year, are expected to be able to recover their fixed costs from the very high capacity payments paid during the short periods of peak demand. If there is a shortage of capacity, owners of peaking plants make large profits because these peak-load payments are activated more frequently than would be the case in an optimum system, and new investment in peaking capacity is stimulated. In order to arrive at a wholesale price for power, the Pool selling price, a number of small payments, for example, to reflect system costs such as maintaining a spinning reserve, are added to the Pool purchase price.

Inevitably, the imposition of transitional contracts on RECs which insulated their cost base from the operation of the Pool has meant that the Pool played no great part in price-setting in the first three years of operation of the new system. The Pool price remains highly volatile and unpredictable, and has proved prone to bidding strategies of dubious propriety by the generating companies.

For example, given a large surplus of capacity, the mechanism intended to encourage the building of peaking plants by causing high Pool prices at peak times has not been tested for this purpose. However, it has proved vulnerable to abuse by generating companies who can withdraw plant from bidding at peak times, narrowing the gap between supply and demand and increasing their peak power payments. Apart from these abuses (now much controlled by the regulator), few expect the mechanism to be effective in maintaining peak plants on the system. The capacity payments are far too vulnerable to unusual weather patterns. A succession of mild winters and a surplus of base-load plant has meant that most existing peaking plant has been retired because costs were not being covered, with the result that the plant mix is far from optimal. However, over the period of operation since the Pool was introduced, this surplus of base- and mid-load plants has meant that the shortage of peaking plant has not led to any overall capacity shortage.

An important point to be made about the way the system operates is that although in principle it is reasonably simple, in practice it has become immensely complicated. Many people started out monitoring Pool prices, full details of which are published. However, it rapidly became clear that keeping up to date with the developments—and more important the political and economic factors which lay behind Pool price setting—was beyond the means of all but the largest companies. Thus, very few specialists and even fewer electricity consumers genuinely understand the mechanics of trading. This, and the dominance over Pool price-setting by National Power and PowerGen, has been an effective deterrent to the emergence of, for example, a futures market in electricity.

Overall, there is no sign that the participants in the new system have any wish for the Pool to become more influential. Few power purchasers are choosing to buy power direct from the Pool or at prices closely linked to Pool prices and, as discussed later, there is little sign that new generation companies will be prepared to build plant which will depend on successfully bidding into the Pool for its profitability. This lack of confidence amongst consumers and producers in the Pool has potentially serious long-term consequences for the future of the Pool, which are examined in detail later.

IV.B. Fuel choice

Most independent commentators predicted that nuclear power and British-mined coal would be heavy losers from the privatisation process,

Table III.2. Fuel input for electricity generation in the UK (mtoe)

	1970	1975	1980	1985	1987	1989	1991	1993	1994
Coal	48	47	56	47	55	52	53	40	37
Oil	13	14	8	11	6	7	8	6	4
Natural gas	0.1	2	0.4	0.5	0.9	0.5	0.6	7	10
Nuclear	7	8	10	17	14	18	17	21	21
Hydro	0.4	0.3	0.3	0.3	0.4	0.4	0.4	0.4	0.4
Total	69	71	75	75	78	79	80	76	75

Source: Digest of United Kingdom Energy Statistics (various years).
Note: A full series of separate demand data is not available for England and Wales.

and so it has proved (see Table III.2). The problems with nuclear power became apparent during privatisation, and forced the major changes noted previously to the structure of the generation side of the business.

IV.B.a. Nuclear power

A strong element contributing to the demise of the original proposals to sell off the nuclear power plant was its cost. At the time of privatisation, far from being the cheapest option available as the CEGB had argued at the Sizewell B Inquiry, the then chairman of the CEGB and the man chosen to be chairman of National Power, Lord Marshall,[12] estimated that the cost of power from the new Sizewell B PWR would be about twice that of competitors such as coal and gas (Lord Marshall, 1989). The main reason for this new estimate of nuclear costs was a change in the rate of return required on capital. At the Sizewell B Inquiry opened in 1981, the government's rate-of-return hurdle which Sizewell had to pass was 5% real *per annum*. This was raised to 8% in 1988, a level sufficient to swing the balance against nuclear power. In the private sector, 12–15% appears to be the target rate of return required, even for low-risk projects. With these higher hurdles, even Nuclear Electric's most optimistic estimates of the costs for a new PWR could not produce electricity prices which would be competitive with those of coal- or gas-fired plant. This change in the discount rate does not only affect nuclear power, but works against any capital intensive option, such as tidal barrages.

Even if the estimated costs per kWh of nuclear power were as low as those of gas, it is unlikely that such a capital-intensive option would be chosen by financiers. Capital intensity leads to a rigidity of costs

---

[12] Lord Marshall's career had mostly been in the nuclear industry, and when nuclear power was withdrawn from privatisation there was little logic in him maintaining the position as Chairman of National Power, and he then resigned.

which, in a competitive market, leaves the plant vulnerable to changes in commodity prices. If market conditions turn against a coal- or gas-fired plant, the operator can usually at least recover most of one of his major costs, that of buying fuel. If market conditions turn against nuclear power plants, nothing can be done to avoid the major cost i.e. that of repaying capital.

However, cost was far from the only element. If cost had been the only problem, a subsidy such as now exists could as easily have been directed to a private as a public company. This is now being demonstrated in the privatisation of the UK railways in which operators of socially desirable services will receive subsidies. To cost must be added the financial risks particular to nuclear power. Such risks are often beyond the control of the operating company, and so a good 'track record'—one of the finance community's criteria for judging promoters of risky projects—does not help much. For example, a nuclear accident anywhere in the world is likely to impose additional costs on all nuclear operators as safety regulators try to reassure the public that a similar event is not possible at a plant under their jurisdiction. The government refused to underwrite this type of largely open-ended and unquantifiable risk, and withdrew nuclear power from privatisation. It was thus the economic risks as much as the high cost of nuclear power and its capital intensity that caused the change in policy (Thomas, 1994a). In the statement announcing the withdrawal of the nuclear sector from privatisation, the Energy Secretary, John Wakeham said that 'unprecedented guarantees were being sought' by the 'electricity industry and investment community' and that he was 'not willing to underwrite the private sector in this way' (Power in Europe, 1989).

The government has completed the review of nuclear power promised when the nuclear sector was withdrawn from privatisation (Department of Trade and Industry and the Scottish Office, 1995). One of the main outcomes of the review was a proposal to reorganise the nuclear sector. All the AGRs, including the two in Scotland and the Sizewell B PWR, were privatised in a single company, called British Energy. The sale was completed in July 1996. The Magnox plants were retained in state ownership in a new company called Magnox Electric, which will eventually be under the control of the state-owned fuel and nuclear services company, British Nuclear Fuels (BNFL). The FFL was be reduced to about 4% at privatisation, with the proceeds of this levy going only to the government via Magnox Electric. The arrangements by which Scottish Nuclear receives a premium price for its output were also removed.

Thus, it seems likely that nuclear power will turn out to have been the first victim of the structural reforms, despite the government's commitment to promote nuclear power. This will clearly be a source of satisfaction to those with ideological objections to nuclear power. However, it seems also to be welcomed by a much larger group of people: those with no objection on principle to nuclear power, but who were concerned that energy policy had been systematically and heavily distorted in favour of nuclear power over the past few decades by the pro-nuclear lobby.

IV.B.b. British coal

The next victim of electricity privatisation is likely to be the British coal industry. Whereas the nuclear industry had the support of the government, this was not the case for the British coal industry. As noted earlier, former government ministers have openly acknowledged their wish to break the power of the largest British miners' union, the NUM. Historically, British coal has been expensive in comparison with internationally traded coal. In recent years, technical innovations and new working practices led to large sustained productivity growth and cost reduction, but these cost reductions for British coal did little more than keep pace with falls in world coal prices (Surrey, 1992). Nevertheless, a strong case could have been made that, given the continuing likelihood of cost-reducing technical change in the British coal industry, over a long period, retaining a British coal industry would have cost electricity consumers no more than the alternatives. However this was not the issue at the forefront of the government's mind.

By the time of the expiry of the government-imposed coal contracts in 1993, the impact of the decisions to build gas-fired power stations was being strongly felt in National Power and PowerGen's demand for coal, and coal stocks at power stations were growing rapidly. More gas-fired plants were due to be completed which had take-or-pay gas supply contracts, which meant they would be operated in preference to coal-fired plants. The generators expected their demand for British coal to be no more than half that at time of privatisation. The negotiations for the renewal of the three year contracts inevitably raised the prospect that the most of the British deep-mined coal capacity would have to close, devastating the communities dependent on these mines. A considerable public protest movement built up, but by then the situation for coal was irretrievable. New five-year contracts were eventually brokered by the government between British Coal and the large generators, matched by supply contracts for differences between the generators and the RECs. The coal contracts were for about half the annual volume of coal covered in the first contract, and prices were set to fall by nearly 30% in real terms.

In 1998, when the market for supply to final consumers was to be fully opened, the RECs would lose their guaranteed market, and electricity supply contracts beyond this point would not have been commercially viable. However, the coverage of the contracts for differences between generators and RECs was sufficient to mean that the Power Pool would not have a significant direct role in setting prices until 1998.

The British coal industry had a fragile existence, being potentially viable largely in only one sector of one geographic market, the UK power sector, where 80% of British coal was sold. It could be competitive with imported coal in this market by off-setting its higher mining costs against the lower transport costs which were incurred, because coal-fired power plants have historically been sited close to the mines which feed them. It was also extremely vulnerable to shifts in the sterling–dollar exchange rate, because internationally traded coal is priced in dollars. If sterling sinks to £1 = $1.10, as it did a few years ago, the British coal industry is highly competitive, but if it rises to £1 = $2, a rate which has also prevailed in recent years, British coal becomes uneconomic. But coal mines cannot be turned on and off at will, and if a pit has to stop producing because its short-term prices are above those of the world market, it will have to be abandoned and the reserves will quickly become irrecoverable.

If a British coal industry was to survive, it would therefore have needed the assurance of long-term contracts for the purchase of its output to give sufficient security for the necessary continuing investments in new mining capacity to be undertaken. The new generating companies were unlikely to want to give this for a number of reasons. First, it introduced rigidity into their cost structure, because if coal is effectively bought on take-or-pay terms, it becomes a fixed cost like capital costs. Second, they wanted to diversify their fuel sources which, at the time of privatisation, were heavily dependent on British coal.

Unlike nuclear power, support for the use of coal for power generation is not usually a matter of basic ideology, but the government had its own agenda with the miners' union. In addition, there were those who believed that the relationship between British Coal and the CEGB was deeply unhealthy because of the lack of competitive pressure exerted by the customer on the supplier. Such people were not unhappy that the British coal industry lost what they saw as the unjustifiably privileged position it enjoyed for 40 years.

British Coal was privatised in 1995 by a trade sale to a privately owned company rather than by a public flotation of shares. Contrary to expectations, the industry was not split into competing units, but the whole of the English mining sector, which includes the majority of production capacity, was sold to a single company (R. J. Budge) for £815 million. The sale appears to have been priced on the basis of the value of the 1993 contracts with National Power and PowerGen, which expire in 1998. There is a strong expectation that the new mine owners will fulfil these contracts at the minimum operating cost with no investment in future mining capacity, and they will have no interest in maintaining the existing mines in an operable condition after 1998. This will leave Britain with very little deep-mined coal production capacity.

It is therefore not hard to explain why nuclear power and British coal are now unattractive as sources of power. In both cases, their capital intensity, the need for long-term contractual commitments and the long lead-times of new power plants introduce extra risks to investors. For nuclear power, there are unquantifiable and potentially open-ended risks and costs, such as those arising from waste disposal, decommissioning and changes in safety regulations. For British coal there is a history of government interference in decisions on its use. But, prior to privatisation, most commentators expected the main 'winner' from the process to be imported coal burnt in existing power stations. While there has been some increase in coal imports, this has been far from the most important effect.

If nuclear and coal were the victims of privatisation, it is natural gas consumption which has been the winner. The unexpected availability of a fuel new to the generating market, natural gas, and a new, efficient technology, the combined-cycle gas turbine (CCGT), to use it has contributed massively to the apparent success of the privatised system. Indeed, the only unsubsidised power plant orders placed since privatisation have been for gas-fired CCGTs.

Combined-cycle technology had been growing in use outside the UK and Europe at a slow pace throughout the 1980s. In Europe, since the first oil crisis in 1973/74, natural gas was seen as a premium fuel too valuable to be used as a crude heat source, and electricity generation using gas was restricted under European Union regulations which were not lifted until 1989. Similar restrictions were in place in the USA. However, the increasing reserves of natural gas worldwide had meant that such a rigid view of natural gas was coming into question well before then and, by the mid- to late 1980s, the removal of the restrictions was widely anticipated.

Combined-cycle technology offered low specific capital costs and much higher thermal efficiencies. It also offered lower emissions of greenhouse and acid gases. This meant that instead of having to retrofit expensive flue-gas desulphurisation (FGD) plant at existing coal-fired stations, National Power and PowerGen could meet their targets for emissions reduction by replacing coal-fired stations with CCGTs. It was also a technology which required few 'user skills', and was therefore attractive to new, inexperienced entrants and also to PowerGen and National Power, who could reduce their cost base by running down their extensive back-up and maintenance capabilities. On-site construction was quick and simple, maintenance was easy, and manufacturers and the insurance market were much more ready to guarantee operating performance than they were for complex steam plant burning a highly variable fuel such as coal. Manning levels were low, and obtaining planning consents did not usually present major problems.

For UK consumers, natural gas was available in easily accessible, sometimes quite small, fields in the North Sea. Given a contract to use the full output of a field over a period (usually 15 years) at a price that was certain to yield a good return, a gas developer would be prepared to supply gas to a CCGT at a price guaranteed to be stable.

However, there was one further element needed to launch what became known as the 'dash for gas': long-term contracts between plant owners and electricity distributors. To combat the generating duopoly, the regulator was willing to allow contracts of 15 years between new generating companies and distributors, which effectively took these plants out of the Pool system. In practice, the generators and distributors were usually linked by ownership ties, with most independent generators controlled by a large stake-holding from an REC. While OFFER is not routinely required to scrutinise contracts between RECs and generators, RECs are required to purchase power 'economically', and if the regulator had believed these practices were not desirable, there is a wide range of measures the DGES could have taken to effectively stop the signing of such contracts. For example, the proportion of power which RECs could purchase from sources in which they have a financial stake (currently 15%) could have been reduced, or the issue of power supply contracts could have been referred to the Monopolies and Mergers Commission, a disruptive process which companies would be eager to avoid.

Whether new CCGTs will provide cheaper power than continuing to operate old coal-fired plant is an issue which has been debated extensively, but it has not been the decisive factor in plant-purchasing decisions. Whatever the virtues of CCGTs, the main motive for buying them was not to reduce the price of power. The existence of contracts and franchises meant that distribution companies did not appear to need to fear the consequences if CCGTs they bought turned out a little more expensive than the alternative (namely purchasing power from National Power or PowerGen), probably based on use of existing power stations burning British coal. The priority for the distribution companies was to reduce the power the duopoly could exert over them (see Table III.3 for a list of CCGTs ordered by companies other than National Power and PowerGen). The market power of the duopoly meant they were similarly insulated if they made choices which raised prices to consumers: their priorities were to absorb the available gas, thus raising the barriers to new competitors, to avoid capital expenditure on FGD plant, and to reduce their dependence on British coal (see Table III.4 for a list of CCGTs built by National Power and PowerGen).

The remarkable burst of orders for CCGTs of about 13 GW of plant,

Table III.3. Combined-cycle power plant owned by new generators

Station	Owner	Size (MWe)	Order date	Commission date	REC partners
Barking	Barking Power	1000	1992	1995	Southern, Eastern, London
Spondon	Derwent	214	199?	1994	Southern
Sellafield	Fellside	170	199?	1994	—
Stallingboro	Humber Power	750	1994	1996	Midland
Keadby	Keadby	680	1992	1995	NORWEB
Kings Lynn	Eastern	340	1994	1996	Eastern
Medway	Medway Power	660	1992	1995	SEEBOARD, Southern
Peterborough	Peterborough	405	1991	1993	Eastern
Corby	Corby Power	350	1990	1993	East Midlands
Greystones	ICI/Enron	1875	1990	1993	Midlands, South Wales, Northern, South Western
Roosecote	Lakeland	220	—	1991	NORWEB
Brigg	Yorkshire	240	1991	1993	Yorkshire
Total		6944			

Table III.4. Combined-cycle power plant owned by National Power and PowerGen*

Station	Size (MWe)	Order date	Commission date
National Power			
Deeside	500	1991	1995
Didcot	1360	1994	1997
Little Barford	680	1992	1996
Killingholme A	620	1990	1994
Total National Power	3160		
PowerGen			
Connah's Quay	1442	1992	1996
Killingholme B	900	1990	1992
Rye House	700	1991	1993
Total PowerGen	3042		

* Includes only stations which are under construction or in service on 1 December, 1994.

equivalent to more than 20% of the entire current capacity of the England and Wales system, is having a dramatic effect on coal consumption. The CCGTs and the Sizewell B PWR, commissioned in 1995, will be operated at as high a load factor as possible. If it is assumed that this new plant substitutes directly for coal-fired plant, coal demand will be 46 million tonnes a year lower than it would have been. This is approximately equal to 60% of annual coal use for power generation in England and Wales prior to privatisation.

An unintended beneficiary of the need to provide a consumer subsidy to nuclear power has been non-conventional power sources. To give the subsidy at least a veneer of respectability, it was presented as a levy on fossil-fuel generation to stimulate non-fossil fuel generation in the cause of reducing the emissions of greenhouse gases. A small proportion of the FFL (so far between 2 and 8% *per annum*) therefore goes to subsidise non-conventional power sources. Much of this subsidy has gone to existing plants and sources such as landfill gas and sewage gas, which are not what are usually thought of as renewable. Only wind-power amongst the genuine renewables has benefited to any extent. Nevertheless, whilst there can be some criticism of how the FFL has been distributed, there can be little doubt that renewables have done better than they would have done had the old organisation remained in place. It is not clear what arrangements will be made to subsidise non-fossil fuel sources other than nuclear power after the phase-out of the FFL.

Energy efficiency was a slow starter under the new regime, with the electricity regulator initially suggesting that encouraging energy efficiency was outside his remit. However, it was successfully argued that the energy industries had a strong financial incentive to maximise their sales, and this was a powerful disincentive to energy efficiency measures. There have been attempts to adjust the regulatory formula for the gas industry to generate funds for an Energy Saving Trust (EST). However, the new Director General of Gas Supply, Claire Spottiswoode, has frustrated attempts to impose such an adjustment on the grounds that it was a form of taxation, and levying taxes exceeded her powers. Whether generating funds in this way will prove possible remains to be seen. The EST has estimated that it requires £1.5 billion between 1993 and 2000 if the UK's CO_2 targets are to be met,[13] but there must be

---

[13] Of the 10 million tonnes of carbon emissions the UK Government projects, 25% is planned to be met by the work of the Energy Saving Trust. See UK Department of Environment (1994, p. 24). In evidence to that Committee, the Energy Savings Trust detailed the timing and size of their investment requirements. See House of Commons Environment Committee (1993, Vol. II, pp. 71–75).

serious scepticism as to whether such a large sum of money will be forthcoming. As a result of OFFER's 1993 review of the RECs' supply businesses, £1 per customer per year has been allocated to expenditure on energy efficiency projects since April 1, 1994, but this will provide only £25 million per annum.

IV.C. Regulation

IV.C.a. Regulatory policy
Whereas the first Director Generals of Telecommunications (Professor Bryan Carsberg) and of gas supply (James McKinnon) were relatively low-profile appointments, the choice of the first Director General of Electricity Supply, Professor Stephen Littlechild, was much more significant. Littlechild, previously Professor of Commerce at the University of Birmingham, had contributed considerably to forming government policy on privatisation. He had written extensively for the Institute of Economic Affairs (IEA), usually seen as a free-market 'think-tank' for the Conservative Party, and had been commissioned by the Government in 1982 to write a study on regulation of the privatised telecommunications industry (Littlechild, 1986). This report argued strongly for local tariff reduction regulation, better known as 'RPI−x' or incentive regulation, rather than, for example, the US model of rate-of-return regulation (by which profits are limited to a specified rate of return on the capital employed). Elsewhere, Littlechild characterised rate-of-return regulation as 'government nannying' implying that using the RPI−x formula and vigorous pro-competition policy would lead to a much lighter form of regulation. Littlechild also argued that privatisation of the electricity supply industry would bring substantial benefits to consumers (Beesley and Littlechild, 1986) and he had written on the form of regulation which would be most appropriate for the electricity supply industry before his appointment as DGES (Beesley and Littlechild, 1989).

Littlechild therefore had a prior commitment to the form and style of regulation of the electricity supply industry, and had a strong self-interest in making a success of it. The government planned that the focus of most regulation would be on the natural monopoly parts of the system and on overseeing transitional arrangements for non-monopoly parts, such as electricity supply, to full competitiveness.

While the RPI−x formula seems a powerful tool to control prices, it must be stressed that much of the cost of power is outside its scope. Thus, a major element in the final cost of power, the generation cost, is not covered by the formula, on the grounds that generation should be a competitive market. RECs are therefore able to pass power purchase costs through to consumers in their tariffs with no formal mechanism to control these costs. Of course, if the generation market is genuinely competitive and the right of consumers to purchase power supplies from any licensed supplier does put heavy pressure on the RECs to minimise their tariffs, then there will be no need to use regulation to control these costs. However, this is not the case at the moment, nor is it certain that the market will develop in this way.

Regulation using the RPI$-x$ formula has also proved more complex than was anticipated. In principle, it would not appear to be difficult to derive reasonable efficiency targets, but special investment needs and other factors have made the process contentious. Indeed, in the most recent review of the application of the formula to the distribution business of the RECs, the DGES decided to depart from the principle of RPI$-x$ price control by imposing a one-off price cut in response to the apparently excessive profits the RECs were making (Office of Electricity Regulation, 1994). Whether OFFER has the resources to scrutinise fully the plans and investment requirements of the companies and to assess the scope for cost-reducing technical change is not clear. If OFFER does not have this capability, the RECs and the NGC will ensure that the regulator will set targets which are easily met and do not exert strong competitive pressure on them.

However, controversy has also surrounded the regulation of the generation companies, an area in which OFFER was not expected to become heavily involved. The basic problems arise from the structure of the generation side, which is effectively controlled by the duopoly position of National Power and PowerGen.

One element of the strategy to weaken the power of the duopoly, already noted, was the decision to allow new generation companies, usually part-owned by RECs, to sign long-term supply contracts with RECs. Such long-term contracts are non-competitive on a number of grounds: first, the element of 'self-dealing' is worrying; second, there has been no explicit competitive process which marks out the projects chosen as the best available; and third, there is no competitive force on these new generation companies other than doing sufficient to supply power at the contracted terms—if a new cheaper source comes along, the generator can maintain its market share without having to respond. Their only competitive virtue is in their contribution to the breaking up of the duopoly of National Power and PowerGen. However, the CCGTs only break the power of the duopoly in the base-load market. In the mid- to peak-load market, currently supplied by older coal- and oil-fired plant, and which for much of the year determines the Pool price, the duopoly's supremacy is unchallenged. Once CCGTs have achieved what they can in breaking the duopoly, there seems little logic in allowing long-term contracts if a fully competitive market is being sought.

IV.C.b. Regulatory style

To illustrate the style and frequency of the regulator's intervention, Table III.5 lists the major interventions and decisions made by the regulator in the calendar year 1994. This shows the extent to which the regulator is being drawn into new areas, such as nuclear policy and energy efficiency measures, because they have some impact on economic regulation, but which are far removed from style of intervention that would have been expected from the government's rhetoric on privatisation.

V. Evaluation and Issues Arising

V.A. The record to date

V.A.a. Supply security and prices

The traditional criteria for evaluating the effectiveness of an electricity supply system are security of supply and affordability. The evidence available so far on these issues has only limited value in forecasting the long-term performance of the system. At the most basic level of keeping the lights on, security does not appear to have been compromised. Indeed, far from leading to a shortage of plant, the first few years of the new system are projected, all other things being equal, to result in a surplus of plant. In practice, a large surplus is unlikely to arise because unneeded plant will be retired when it becomes surplus. However, given the importance of electricity supply to the economy, there was never any risk that the government would allow the new system not to work at this most basic level. In addition, the new companies, National Power and PowerGen, had a strong incentive not to cause problems as this might have jeopardised their future, especially if there had been a change of government.

In terms of cost to consumers, UK prices appear to be in the middle of the range for European countries (see Table III.6). These comparisons must be treated with great care as they are vulnerable to a number of possible distortions: for example, fluctuations in currency exchange rates may change relative prices and they may also be distorted by differences in the price review cycle, for example, a utility which has just raised its prices is likely to be unfairly treated if it is compared to a utility which is about to raise its prices.

Since privatisation, there does not seem to have been any dramatic

Table III.5. Major decisions by the DGES in 1994*

Date	Decision
February[1]	Decision not to refer National Power and PowerGen to the Monopolies and Mergers Commission. In return, National Power and PowerGen would attempt to sell 4000 MW and 2000 MW, respectively, of their coal-fired plant in order to increase competition. National Power and PowerGen also agreed to bid their plant in such a way that the Pool price would average 2.4p/kWh for the following two years, a reduction of about 7%.
March[2]	Consultation exercise launched to determine whether trading outside the Pool should be allowed and whether the Pool pricing mechanism should be reformed.
April[3]	Efficiency standards review published giving RECs target demand savings. Cost of the schemes to be financed by a levy of £1 per customer.
June[4]	Yield and distribution of FFL announced.
July[5]	Results of March consultation announced. Trading outside the Pool not to be allowed and no reforms to the Pool pricing mechanism recommended.
August[6]	Results of review of RECs' distribution business announced. One-off price cut of 11–17% (size of cut varied from REC to REC) and 'x' of RPI $-x$ formula set at 2% until 1999/2000. Additional energy efficiency incentives for RECs introduced. Standards of service RECs must meet raised. RECs accept the review.
September[7]	Annual report on REC customer service published.
October[8]	OFFER submission to the Nuclear Review published recommending privatisation of the two UK nuclear generators, a redistribution of assets to allow competition between them, an end to the FFL and no Government subsidies for the construction of new nuclear plant.
November[9]	Scottish Hydro rejects OFFER's price control proposals and OFFER refers the issue to the Monopolies and Mergers Commission.
November[10]	Conference speech by Littlechild stating he had insisted on sale of RECs' share-holding in the NGC. Also advocated sale of NGC's pumped storage plants and separation of RECs' monopoly distribution business from other activities to avoid anti-competitive behaviour.
November[11]	OFFER report published critical of the allocation of the non-nuclear NFFO funds.

Sources: [1] Reported in *Power in Europe* (1994) No. 167, pp. 1–2. [2] Reported in *Power in Europe* (1994) No. 169, pp. 2–3. [3] Reported in *Power in Europe* (1994) No. 171, p. 24. [4] Reported in *Power in Europe* (1994) No. 176, p. 22. [5] Office of Electricity Regulation (1994a). [6] Office of Electricity Regulation (1994b). [7] Office of Electricity Regulation (1994c). [8] Office of Electricity Regulation (1994d). [9] Reported in *Power in Europe* (1994) No. 187, p. 23. [10] Address to the Institute of Economic Affairs, November 1994, reported in *Power in Europe* (1994) No. 187, pp. 22–23. [11] Letter from Professor Littlechild to the Energy Minister, Tim Eggar, reported in (1994) *Electrical Review* (1994) No. 23, p. 4.
* Includes only stations which are under construction or in service on 1 December, 1994.

Table III.6. International comparison of electricity prices (p/kWh)

Country	Utility	Domestic*		Industrial	
		1993	1994	1993	1994
England/Wales	North	9.29	9.01	5.14	5.14
England/Wales	Central	9.36	9.27	5.93	5.87
England/Wales	South	8.81	8.52	5.28	5.11
Austria	EVN	11.28	10.78	6.56	6.27
Belgium	National	13.00	12.49	5.79	5.40
Denmark	SEAS	12.05	10.85	4.21	3.75
Finland	Helsinki	6.08	5.66	3.53	3.21
France	EDF	11.72	11.43	4.84	4.64
Germany†	—	12.59	12.48	8.06	7.75
Greece	PPC	7.68	6.77	5.04	4.45
Ireland‡	ESB	8.91	7.90	5.26	4.67
Italy	ENEL	9.43	10.86	6.68	6.55
Netherlands	Average	8.28	8.14	4.54	4.42
Portugal	EDP	12.17	10.66	7.60	6.33
Spain	National	12.75	10.84	6.72	5.51
Sweden	Vattenfall	7.78	7.05	2.89	2.59
Scotland	—	8.13	8.36	5.29	5.29
Northern Ireland	—	9.36	9.59	5.39	5.42

Source: Electricity Association (1994).

* Data for domestic consumers assume usage of 3300 kWh/year, while industrial demand is calculated assuming a maximum demand of 2500 kW and a 40% load factor.

Note: Prices include non-recoverable taxes, but exclude VAT.

† Data for Germany are an arithmetic average of the data given in the study for north, west and south-west Germany.

‡ Data for Ireland refer to urban consumers.

movement in prices (see Table III.7). Some groups of consumers have gained a little, some have lost a little, but these are only minor realignments of costs. The clearest losers are the large industrial consumers which had previously enjoyed arrangements to purchase power at well below economic rates. The formal procedures of the Pool system mean that such subsidies cannot be sustained. However, the issue is more complex than whether real prices have gone up or down. Two questions should be asked: are prices higher or lower than they would have been had the old system remained in place?; and to what extent are the price changes the result of competitive forces and tough regulation rather than the conditions imposed by government at and after privatisation?

It is impossible to determine what decisions would have been taken under the old regime and thus calculate what prices would have been. Nevertheless, there has been a substantial fall in real fuel prices, in particular that of British coal, severe reductions in manpower, and a

Table III.7. Electricity prices in England and Wales (£)

	1990/91 typical bill	1994/95 typical bill	Total increase (%)
Domestic	262.25	286.45	9.2
Commercial	12,409	13,763	10.9
Industrial	150,504	165,762	10.1

Source: Centre for the Study of Regulated Industries (1994).
Notes: Bills are expressed in money of the day (i.e. not inflation-adjusted), and the figures shown represent a simple arithmetic average over the 12 RECs. Domestic consumers are assumed to consume 3300 kWh. For commercial consumers, it is assumed the maximum demand is 100 kW, load factor is 20%, and demand is 175,200 kWh. For industrial consumers, it is assumed the maximum demand is 1000 kW, load factor is 30%, and demand is 2,628,000 kWh.

dramatic reduction in electricity-industry sponsored R&D. The cost of nuclear generation has also fallen markedly, due to reductions in manning and increased output from existing stations. Given this background, the relative stability in prices becomes less impressive. Indeed, Yarrow (1992) has estimated that prices are some 15% higher than they would have been under the old structure. The rent from lower fuel prices and reduced manpower costs appears to have been absorbed in higher profits and dividends, and, to a degree, higher salaries, particularly those for senior management. The Centre for the Study of Regulated Industries (1994) found that, since privatisation, distribution charges had gone up by 14% in real terms, and the contribution of distribution to the average final domestic bill had risen by over 15% to nearly 34%. There is however, variation between RECs, for example, a typical bill for NORWEB domestic consumers increased by only 3%, while for Eastern Electricity consumers, the bill would have gone up by 14.2%.

However, many of these price changes were determined at the time of privatisation by the government. All consumers pay the FFL, which has remained largely unchanged as a 10% surcharge on all bills. The charges for transmission, distribution and supply were set for 3–5 years forward by the regulatory formula determined then. The formula setting the distribution charges, typically 25% of a residential consumer's bill, applied up to March 1995, the formula setting the supply charges (6% of a typical bill) applied up to March 1994, and the formula setting the transmission charges (5% of a typical bill) applied up to March 1993. The new formulae have been more demanding on the suppliers than previously, but they have had little time to have an impact on prices. The new transmission formula requires annual price reductions of 3%, and the supply formula 2%. By March 1995, these targets would only have reduced consumers' bills by about 0.4%. The new distribution formula required one-off price reductions for distribution of between 11 and 17%, as well as annual reductions of 2%. In the first year, this would have reduced overall bills by about 3–5%.

Clearly, the first regulatory formulae could have been imposed on the old system, and so the impact of market forces on the transmission, distribution and supply elements of consumers' bills has therefore so far been very small. However, generation, typically 54% of residential consumers' bills, should be competitive, and it might seem that there was greater scope for market forces to apply here. However, the government imposed coal-supply contracts on National Power and PowerGen for three years, and matching power-supply contracts on the RECs with National Power and PowerGen. These contracts therefore largely determined the cost of generation for the first three years. In 1993, the government had a direct hand in negotiating the new five-year contracts between the large generators and British Coal, and between the large generators and the RECs, so even after 1993, the market was not the primary influence on decision-making.

Overall, the transitional conditions imposed by the government, which could equally have been imposed on the old system, have been the major determinant of price movements so far. Market forces and disciplines have had little opportunity to have an impact yet.

V.A.b. Profits, employment and R&D

The changes in the structure of the power station stock resulting from the 'dash for gas' and the consequent reduction in coal capacity have been the most visible effects of privatisation. However, in other areas, changes which are potentially equally important in the long term have also occurred. Table III.8 shows the financial results of the CEGB and its daughter companies in the years immediately before and after privatisation. Profitability of the privatised companies is now more than 150% higher than the level of CEGB profits a decade ago.

These results must be interpreted with care. The accounts of the CEGB were distorted in 1987/88 and 1988/89 by massive increases in the accounting provisions made for long-term nuclear liabilities, which were necessary if the nuclear sector was to be privatised. In 1989/90, the accounts of the four daughters of the CEGB are somewhat notional, as the companies were being run from within the CEGB as separate divisions and no FFL was raised. For Nuclear Electric, it can be seen that 42–54% of its turnover came from the proceeds of the FFL. Its turnover increased, at the expense of National Power and PowerGen, as a result of the much higher levels of output its plants have achieved— under the ENOR, it could sell all the output it could produce. However,

Table III.8. Financial results of the CEGB and daughter companies (£million)

Company		85/86	86/87	87/88	88/89	89/90	90/91	91/92	92/93	93/94	94/95
CEGB	a	8015	8156	8325	8935	—	—	—	—	—	—
	b	783	675	457	−3026	—	—	—	—	—	—
National Power	a					3998	4378	4701	4348	3641	3953
	b					178	479	514	580	677	705
PowGen	a					2608	2651	3009	3188	2932	2885
	b					234	272	359	425	476	545
Nuclear Electricity	a					2058	2202	2432	2706	2962	2889
	b					−928	−14	62	109	392	1068
	c					0	1195	1265	1280	1230	1251
NGC	a					1071	1144	1320	1392	1425	1428
	b					429	386	498	533	580	611
Total	a	8015	8156	8325	8935	9735	10,375	11,462	11,634	10,960	11,155
	b	783	675	457	−3026	−87	1123	1433	1647	2125	2929
b/a		0.098	0.083	0.055	−0.339	−0.01	0.108	0.125	0.142	0.194	0.262
b/a excluding Nuclear Electric				—	—	0.110	0.139	0.152	0.172	0.217	0.226

Source: Annual reports and accounts of the companies.
Notes: a = turnover in current money, b = pre-tax profit, c = fossil fuel levy (FFL) income. For Nuclear Electric, profits are calculated after the inclusion of the FFL. Accounts are calculated on a historic cost-accounting basis.

as the FFL was only payable on a specified amount of output (40 TWh) a level which it exceeded by more than 50%, in 1995 receipts from the FFL did not increase.

In the three years from Vesting Day (April 1, 1990), the scope for National Power and PowerGen to influence their financial performance was constrained by a number of factors. The three-year contracts for coal and with the RECs which were imposed on them largely fixed their main cost and income items. With Nuclear Electric and the independent producers taking some of their market share, their turnover inevitably tended to fall. The increased level of profitability National Power and PowerGen have experienced is therefore explained mainly by lower costs, for example in fuel purchasing, staff and R&D (see later). Now that the generators have signed new contracts for coal at much lower tonnages and prices, and have also negotiated new contracts with the RECs, the scope for changes in performance will be somewhat greater in the next few years. The profitability of the National Grid Company, which it should be noted was owned by the RECs until 1995, is remarkably high.

Variations in the financial performance of the RECs were also rather restricted because their market share was largely fixed, and contracts with National Power and PowerGen determined their main cost (see Table III.9). The level of profitability of the RECs is not as high as for the generation sector, and some of the increase in profits was accounted for by the RECs' ownership of the NGC. Nevertheless, if the RECs could beat the 'RPI $-x$' formula by reducing staff levels and introducing more efficient practices, they could improve their profitability. Their performance has also depended on the success of any diversification measures, for example into overseas markets. There is some variation between RECs, but most are substantially more profitable than they were prior to privatisation and, overall, profitability has nearly doubled. There does not seem to be any systematic relationship between the profitability of a REC and the extent to which it has increased its prices.

Clues as to how this general increase in profitability has been achieved come from the figures for number of employees (Table III.10) and R&D expenditure (Table III.11). While in the period up to privatisation employment levels in the generation and transmission sector were largely stable, since privatisation staffing levels have been reduced by 50%. This is only in part explained by the growth in market share of independent power projects. There have been strong indications that these job cuts will continue, notably in the nuclear companies and the National Grid Company, areas where cuts have not so far been so heavy. Also, as National Power and PowerGen replace labour-intensive coal plants with CCGTs, employment will fall further in these companies.

Table III.9. Financial results of the regional electricity companies (£million)

Company		1985/86	1989/90	1993/94	1994/95
NORWEB	Turnover	986	1232	1470	1510
	Profits	54	76	178	205
East Midlands	Turnover	987	1263	1445	1370
	Profits	48	91	51	214
MANWEB	Turnover	706	887	920	878
	Profits	26	38	126	86
Yorkshire	Turnover	995	1258	1308	1459
	Profits	55	110	149	217
Southern	Turnover	1125	1457	1780	1680
	Profits	69	128	222	202
South Wales	Turnover	493	604	605	642
	Profits	26	26	104	123
SEEBOARD	Turnover	803	902	1220	1196
	Profits	50	58	132	142
London	Turnover	927	1148	1310	1209
	Profits	85	126	187	172
Eastern	Turnover	1271	1616	1850	2061
	Profits	89	120	177	203
Northern	Turnover	660	820	1030	1081
	Profits	47	66	128	141
South Western	Turnover	601	748	900	875
	Profits	47	66	117	112
Midlands	Turnover	1041	1295	1420	1457
	Profits	66	85	195	178
All RECs	Turnover	10,594	13,310	15,257	15,417
	Profits	662	990	1766	1995
	Profits/Turnover	0.062	0.074	0.116	0.129

Source: Annual reports and accounts of the companies.
Notes: Turnover is expressed in current money, profits pre-tax, and accounts are calculated on a historic cost-accounting basis.

Job losses in the RECs have not been so severe, only amounting to about 20%, but there have been a number of announcements of further significant redundancies for the next few years, perhaps in expectation of greater competition as more consumers are free to choose their supplier. The wave of takeover bids for RECs by foreign utilities, generating companies, water companies and multinationals is likely to lead to a fresh round of redundancies as the new owners look to reduce costs so that the high prices being bid for the companies can be justified.

For R&D, the picture is even more spectacular, particularly with Nuclear Electric. The sharply rising trend in R&D expenditure of the mid-1980s has been reversed and, with most nuclear R&D budgets fixed for several years forward, the figures will decline even further as historic

Table III.10. Employment in the electricity supply industry

	85/86	86/87	87/88	88/89	89/90	90/91	91/92	92/93	93/94	94/95
CEGB	48,274	48,040	47,930	47,631	—	—	—	—	—	—
National Power	—	—	—	—	16,977	15,713	13,277	9934	6955	5447
PowerGen	—	—	—	—	9430	8840	7771	5715	4782	4171
Nuclear Electricity	—	—	—	—	14,415	13,924	13,300	12,283	10,728	9426
NGC	—	—	—	—	6442	6550	6217	5666	5127	4871
All CEGB	48,274	48,040	47,930	47,631	47,264	45,027	40,565	33,598	27,592	23,915
RECs	83,527	82,586	82,212	82,291	82,485	82,288	81,135	77,329	71,673	65,787
All ESI	131,801	130,626	130,142	129,922	129,749	127,315	121,700	110,927	99,265	89,702

Source: Annual reports and accounts of the companies.
Notes: Employment is measured as the average number of employees during the financial year.

Table III.11. Research and development by the CEGB and its successors (£ million)

	85/86	86/87	87/88	88/89	89/90	90/91	91/92	92/93	93/94	94/95
CEGB	135	162	199	201	—	—	—	—	—	—
National Power	—	—	—	—	22	26	17	20	24	—
PowerGen	—	—	—	—	5	14	12	10	9	—
Nuclear Electricity	—	—	—	—	116	95	71	64	51	—
NGC	—	—	—	—	7	7	8	8	8	—
Total	135	162	199	201	150	142	108	102	92	—

Source: Annual reports and accounts of the companies.
Notes: No figure for R&D was reported for the NGC in 1989/90, and the figure is imputed.

commitments in this area are completed and not renewed or replaced. R&D carried out by the RECs is difficult to estimate due to incomplete reporting, but has typically been about an order of magnitude lower than for the generation sector, and reducing spending is therefore much less of a priority for the RECs.

V.B. Issues to be faced

V.B.a. Is light regulation credible?
The political rhetoric of privatisation demanded that regulation be 'light'. The idea that extensive regulation would be required to keep the behaviour of private sector companies in check would have been damaging to the credibility of privatisation. Nevertheless, the forecast that regulation could be light and would quickly wither away as competition took over has always seemed unrealistic. The Pool has proved not to be the centre point of the system and power-supply contracts (which have minimising risk to financiers as a much higher priority than ensuring competition) dominate. Clearly a regulator obliged to promote competition cannot stand by in such circumstances. The issue of whether an electricity supply system centred on a highly competitive Power Pool is feasible.

For the parts of the industry which it was acknowledged would have to be regulated for the foreseeable future, i.e. the natural monopoly areas, the outlook seems to be no more encouraging. Companies with the size and resources of the NGC and the RECs inevitably try to ensure that the efficiency targets imposed on them are readily met, either giving them an easy life or additional profits to shareholders, and the high level of profitability of the regulated sectors suggests they have so far been successful in this. The regulatory body needs to have sufficient resources and capabilities to counter these pressures if it is to be effective.

V.B.b. What to do with nuclear?

The creation of a publicly owned, heavily subsidised nuclear power company was an anomaly to the system that a Conservative government was eager to remove at the first opportunity. The government review of nuclear power, which was completed belatedly in May 1995, confirmed this objective. However, privatising at least some of the nuclear industry was not easily achieved, and the guarantees the City demanded significantly reduced the exposure of the sector to market forces.

In many respects, the record of Nuclear Electric in the period since its creation was impressive. Costs and performance at the operating plants have improved significantly, and the Sizewell B PWR was completed largely to schedule and with limited cost escalation in the final years of its construction.

Nevertheless, Nuclear Electric was still in receipt of the FFL, which comprised about 40% of its income. The subsidy was so large that Nuclear Electric made a substantial cash surplus which was mostly deposited in the Treasury's national loans fund. The revenue from the FFL was unassigned income to Nuclear Electric and, contrary to some assertions, was not specially ear-marked to pay for waste disposal and decommissioning. This has meant that Nuclear Electric, despite being technically insolvent, was able to use the proceeds of the FFL to finance the construction of Sizewell B, avoiding any recourse to borrowing.

This, and the reductions in operating costs of the AGRs and Sizewell B, mean that a privatised nuclear company can probably expect to make adequate provision for their decommissioning and still make an acceptable level of profit in the private sector without need for further subsidy. The substantial financial burdens of decommissioning and reprocessing that the Magnox plants entail will now be shared between the electricity consumer and the taxpayer.

V.C. The long term

The question for the long term can be simply put. Is a privately owned, fully competitive structure a good way to run an electricity supply system? The relatively short period since privatisation has given some clues about what the major issues to be considered might be.

V.C.a. Supply security

The first few years of operation of the new system have seen substantially more plant ordered than is needed for supply security reasons. However, as argued previously, the 'dash for gas' was triggered by a number of special factors. These included the desire of the RECs to reduce the duopoly power which the two main private-sector generating companies held over them, and the wish of National Power and PowerGen to reduce their dependence on British coal. Even so, the 'dash for gas' was only possible because of the willingness of the regulator to tolerate long-term contracts between RECs and new generation companies, and because the duopoly power of National Power and PowerGen minimised the risk of investment in new plant not being recovered. However, once the transitional stage has passed and the duopoly power of National Power and PowerGen has been broken in the base-load segment of the market, it will be hard to justify allowing long-term contracts if a fully competitive generation sector is being sought. In these circumstances, a shortage of plant could as easily result as an excess, illustrating a key weakness of the new system: there are no planning procedures which have the objective of balancing supply and demand for power plants. However, the problem may go deeper than that. In a fully competitive electricity supply system, building a power plant may be too great a risk for the commercial market to take on.

Without a long-term contract for the power, a generating company building a plant would be forced to rely on successful bidding into the Pool or obtaining short-term contracts at Pool-related prices. When it planned a plant, it would not only not know what price to plan on receiving, it would not even know whether the plant would be used at all, and long-term fuel supply contracts could not be justified. It might still be possible to finance new power plants, but they would carry a hefty (perhaps prohibitive) risk premium on the cost of capital to reflect the uncertainty of the Pool. These extra costs would have to be passed on to the consumer, and would also give a strong advantage to large generators who would not face such large risks because of their diverse portfolio of plants. If the principle of a competitive Pool determining the price of power is to be followed and supply is to be secure, the future would appear to hold the risk of either prohibitively expensive power because of the risk premium applied to building new capacity or a continuation of the National Power/PowerGen duopoly (Thomas, 1994b).

V.C.b. How can strategic fuel and technology decisions be taken?

One of the *de facto* duties of the old-style nationalised industries was to carry the national interest in a range of areas. Equipment-supply companies were supported, local fuel sources could be encouraged, and utility research departments could be used to develop technology for the benefit of the nation. This role disappeared overnight with privatisation—equipment was procured on the open market, with few of the early plant orders going to British suppliers; similarly, contracting arrangements which gave a long-term future to British Coal could not be negotiated, and support for research projects such as fast-breeder reactors was withdrawn at the first opportunity.

In some respects, this is a loss which need not be mourned. Too often, what started out as strategic support for important national capabilities became expedient means of supporting uneconomic and uncompetitive but politically powerful groups. In addition, in the context of a European Union Single European Market and GATT (now the World Trade Organisation) negotiations, maintaining strategic national capabilities seems increasingly to be no more than national chauvinism. However, the inability to take strategic decisions may still be regretted.

British Coal's market share did not collapse because British coal was more expensive than imported coal or because power generated using it was more expensive than new CCGTs: it was not. It collapsed because the companies had strategic reasons of their own for investing in gas-fired plant and because the generators were uncomfortable dealing with a state-owned company with heavy political baggage, self-evidently not the best set of reasons for losing a major national resource for ever.

On equipment supply, the break-up of the existing electricity supply industry structure and the introduction of competition in the UK and Europe has not only resulted in individually weaker electricity companies, it has removed the incentives for them to co-operate in technology development. This comes at a time when a major concentration of the equipment supply industry, leaving only three major groupings in Europe with only two further groupings in the rest of the world, has taken place. For an industry which could yet concentrate further, and which has a dubious record on cartelisation and price-fixing, this hardly suggests there will be vigorous price and technology competition for equipment supply in the long term. Nor does it suggest that the voice of the user will be strongly heard in the design and development of new equipment.

The loss of 'user skills' which the run-down of research and support departments has led to may also be regretted. If gas is not an option when new capacity is needed, the industry will have to go to options (be they nuclear power, integrated gasification combined cycle or fluidised bed combustion) which have high capital costs, long lead-times and which require the sort of user skills which the UK industry has been quick to shed. It is far from certain that the private sector will be willing to recreate these capabilities unless there are strong guarantees that this investment in capabilities is sure to be reimbursed.

V.C.c. How will the new companies develop?
Perhaps the most difficult area to examine is the changes which will happen when the newly privatised companies reach maturity. While those in the new companies have often been keen to adopt the style of private industry, portraying themselves as tough, profit-driven businessmen, the companies outwardly remained reassuringly unchanged. The Government held a 'golden share' in the RECs until 31 March, 1995, which gave them effective power of veto over any policies of which they disapproved. For most of the period from privatisation until then, the RECs still appeared as locally based companies whose prime business was delivering power to a specific region.

However, with the publication in August 1994 of OFFER's proposals for a new regulatory formula which would apply to distribution costs from March 1995 (Office of Electricity Regulation, 1994) and the imminent end to the golden share the RECs began to be seen as attractive investments for corporate raiders. OFFER's proposals represented a significant tightening of controls. Previously, the regulatory formula allowed prices to rise between 0.25 and 2.5% per year, varying from REC to REC, but the new formula imposed annual reductions of 2% for all RECs together with a one-off price reduction of 11–17% (varying from REC to REC).

Financial analysts judged that these requirements would be easily met, despite the protestations the RECs had made in the negotiations leading up to the publication of OFFER's proposals, and the share price of the RECs rose dramatically. The weakness of OFFER's proposals appeared to be confirmed when a diversified British company, Trafalgar House, put in a bid for the REC Northern Electric in December 1994. The bid was rejected by the Northern Electric board and the defence it put up revealed the very high profits it expected to make under the new price controls. The DGES found a pretext for withdrawing his proposals despite the fact they had already been accepted by the RECs, and in July 1995, new controls were announced which included a further one-off price cut in March 1996 of between 10 and 13% and an increase in the annual price reductions from 2 to 3% for the three remaining years. This was not judged by the market as being sufficiently stringent to make the ownership of an REC unattractive, and a remarkable wave of takeover bids ensued in the summer and autumn of 1995.

The situation regarding the ownership of the RECs is changing rapidly. By the spring of 1996, nine of the 12 RECs had been subject to proposals for merger or takeover. Some of these have failed, some are being resisted, while others have been accepted. These bids can be divided into five categories, each of which seems to raise important policy issues. While the current bids in these categories may not go through, there is no reason to believe that further bids in these categories will not occur, requiring examination of the issues. The categories are: mergers, takeovers by foreign electricity companies, takeovers by diversified multinational companies, horizontal integrations into other utility industries, and vertical integrations into generation.

There have been serious discussions between RECs of the possibility of mergers. If there was to be a significant reduction in the number of RECs, the scope for yardstick regulation of distribution would diminish, and the competitive field from which consumers could choose to buy their power would shrink. A number of bids by foreign electric utilities have been mooted, and these raise issues about whether there are risks attaching to the loss of national control of the industry. Bids have also been placed by diversified multinational companies, and these bring forward similar problems. They may also make the companies more difficult to regulate effectively unless very stringent rules are applied on the separation of business accounts.

Horizontal integration has also been proposed, with bids by water companies for RECs. This may lead to a risk of cross-subsidisation of a competitive business (electricity supply) at the expense of a monopoly business (water supply). Gas-supply companies may also be interested in diversifying into electricity supply. Perhaps the most worrying set of bids are those which would lead to an integration of generation, distribution and supply. PowerGen, National Power, and the larger of the two Scottish generation companies, Scottish Power, have had bids for RECs accepted. The REC Eastern Electricity now has a substantial generation business, with interests in a number of CCGT plants. It has also acquired some old coal-fired plant from National Power and PowerGen which the DGES required them to sell to reduce their duopoly power. It appeared that the government would not intervene to assess the implications of these bids, despite recommendations to the contrary from the DGES. However, the Trade and Industry Minister, Ian Lang, referred the PowerGen and National Power bids to the government's Monopolies and Mergers Commission (MMC) in November 1995. In April 1996, against the advice of the MMC, Lang ruled against National Power and PowerGen, not allowing their take-over of RECs. However, this decision was one of timing, not of principle, leaving the door open to renewed bids at some later date.

While the pace of takeover activity took many people by surprise, it was probably inevitable that the ownership of the electricity companies would not remain static. If electricity supply is to become an economic activity as much like all other economic activities as possible, there is no reason to assume that the firms involved will wish to remain regionally bound electricity industry companies. Most people in the electricity industry probably believed that the effect of privatisation would be to free them of government interference and, more cynically, allow them higher wages and better fringe benefits. These changes have certainly occurred, but they have by no means been the most significant. The reality for the companies is that their priority must be to seek whatever business opportunities offer the best returns to their shareholders, in whatever country and sector they arise. They have no responsibility to keep the lights on in Britain, much less to promote the cause of the British economy.

In practical terms, this means that there is likely to be some change of ownership of these companies and perhaps even foreign takeovers. Dutch experience with distribution companies and recent UK experience with television franchises suggest that there is likely to be some merger/takeover activity amongst the RECs. There appears to be scope for horizontal integration into other delivered services, particularly the gas market. The companies are also likely to become more international in scope. Whether the public's resulting loss of strategic control and accountability is the business equivalent of a child taking the stabilisers off is bicycle, no loss of safety and much greater potential performance remains to be seen.

Literature

Barnes, M. (1990) *The Hinkley Point Public Inquiries: a Report by Michael Barnes QC*, HMSO, London.

Beesley, M. and Littlechild, S. (1986) Privatisation: principles, problems and priorities. in Kay, J., Mayer, C. and Thompson, D. (eds) *Privatisation and Regulation: The UK Experience*, Clarendon Press, Oxford.

Beesley, M. and Littlechild, S. (1989) The regulation of privatised monopolies in the United Kingdom, *RAND Journal of Economics*, Vol. 20, No. 3, pp. 454–472.

Central Electricity Generating Board (1988) *Hinkley Point 'C' Power Station Public Enquiry, Statement of Case.*

Centre for the Study of Regulated Industries (1994) *The UK Electricity Industry*, London.

Chesshire, J. H. (1989) *Was Britain's Electricity Privatisation Cleverly Done?*, Paper presented to the Financial Times World Electricity Conference, 16–17 November, 1989, London.

Department of Energy (1988) *Privatising Electricity: The Governments' Proposals for the Privatisation of the Electricity Supply Industry in England and Wales*, Cm 322, HMSO, London.

Department of Trade and Industry and the Scottish Office (1995) *The Prospects for Nuclear Power in the UK: Conclusions of the Government's Nuclear Review*, Cm 2860, HMSO, London.

Digest of United Kingdom Energy Statistics (various years) HMSO, London.

Electrical Review (1994) Littlechild attacked over NFFO advice, Electrical Review, Vol. 227, No. 23.

Electricity Association (1994) *International Electricity Prices*, Electricity Association, London.

Hannah, L. (1979) *Electricity before Nationalisation: a Study of the Development of Electricity Supply in Britain to 1948*, Macmillan, London.

Hannah, L. (1982) *Engineers, Managers and Politicians: the First Fifteen Years of Nationalised Electricity Supply in Britain*, Macmillan, London.

Henderson, P. D. (1977) Two British Errors: Their Probable Size and Some Possible Lessons, Inaugural lecture, University College, London, 24 May, 1976, Oxford Economic Papers, Oxford.

Holmes, A., Chesshire, J. and Thomas, S. (1987) *Power on the Market: Strategies for Privatising the UK Electricity Industry*, Financial Times Business Information, London.

House of Commons Environment Committee (1993) *Fourth Report: Energy Efficiency in Buildings*, HC 648-II, HMSO, London.

International Atomic Energy Agency (1986) *Operating Experience with Nuclear Power Stations in Member States in 1985*, International Atomic Energy Agency, Vienna.

Kay, J., Mayer, C. and Thompson, D. (eds) (1986) *Privatisation and Regulation—the UK Experience*, Clarendon Press, Oxford.

Littlechild, S. (1986) *Regulation of British Telecommunications Profitability*, HMSO, London.

Lawson, N. (1992) *The View from No. 11*, Bantam Press, ????.

Lord Marshall (1989) *The Future for Nuclear Power*, The British Nuclear Energy Society Annual Lecture, London, 30 November, 1989.

Monopolies and Mergers Commission (1981) *Central Electricity Generating Board. A Report on the Operation by the Board of the System for the Generation and Supply of Electricity in Bulk*, HMSO, London.

National Economic Development Council (1970) *Report of the Working Party on Large Industrial Construction Sites*, HMSO, London.

Office of Electricity Regulation (1994a) *Report on Trading Outside the Pool*, OFFER, Birmingham.

Office of Electricity Regulation (1994b) *The Distribution Price Control: Proposals*, OFFER, Birmingham.

Office of Electricity Regulation (1994c) *Report on Customer Services*, OFFER, Birmingham.

Office of Electricity Regulation (1994d) *Submission to the Nuclear Review*, OFFER, Birmingham.

Parker, M. and Surrey, A. J. (1992) *Unequal Treatment: British Policies for Coal and Nuclear Power*, 1979–82, Science Policy Research Unit, Brighton.

Parkinson, C. (1992) *Right at the Centre*, Weidenfeld and Nicolson.

Power in Europe (1989) UK nuclear power: the charade is over, *Power in Europe*, No. 62, 9 November, 1989.

Power in Europe (1994) No. 167.

Power in Europe (1994) No. 169.

Power in Europe (1994) No. 171.

Power in Europe (1994) No. 176.

Power in Europe (1994) No. 187.

Report of the Committee of Enquiry into Delays in the Commissioning of CEGB Power Stations (1969) Cmnd 3960, HMSO, London.

Ridley, N. (1991) *My Style of Government*, Hutchison.

Rush, H. J., MacKerron, G. S. and Surrey, A. J. (1977) The advanced gas-cooled reactor: a case study in reactor choice, *Energy Policy*, Vol. 5, No. 2, pp. 95–105.

Scottish Office (1988) *Privatisation of the Scottish Electricity Industry*, Cm 327, HMSO, London.

Select Commission on Energy (1990) *Fourth Report, Session 1989–90*, HC 205-I, HMSO, London.

Select Committee on Energy (1981) *First Report: The Government's Statement on the New Nuclear Power Programme*, HMSO, London.

Select Committee on Energy (1988) *Third Report, Session 1987–88*, HC 307-1, HMSO, London.

Select Committee on Science and Technology (1974).

Surrey, A. J. (1992) Technical change and productivity growth in the British coal industry, 1974–1990, *Technovation*, Vol. 12, No. 1.

The Electricity Council (1987) *Electricity Supply in the United Kingdom: a Chronology—from the beginnings of the Industry to 31 December 1985*, The Electricity Council, London.

Thomas, S. (1994a) A taxpayer's nightmare—shouldering the burden of the 'Nuclear Blight'. In *The Moratorium on Nuclear Power Plant Construction in the UK*, The Power Trust, Taunton.

Thomas, S. (1994b) Will the UK power pool keep the lights on? *Energy Policy*, Vol. 22, No. 8, pp. 643–647.

Thomas, S. and McGowan, F. (1990) *The World Market for Heavy Electrical Equipment*, Nuclear Engineering International Special Publications, Sutton.

UK Department of Environment (1994) *The Government's Response to the Fourth Report from the House of Commons Select Committee on the Environment: Energy Efficiency in Buildings*, Cm 2453, HMSO, London.

Vickers, J. and Yarrow, G. (1988) *Privatisation: An Economic Analysis*, MIT Press, Cambridge, MA.

Walker, P. (1991) *Staying Power*, Bloomsbury.

Williams, R. (1980) *The Nuclear Power Decisions*, Croom Helm, London.

Yarrow, G. (1992) British electricity prices since privatisation, *Studies in Regulation*, No. 1, Oxford Regulatory Policy Institute, Oxford.

Chapter IV
The Norwegian, Swedish and Finnish Reforms: Competitive Public Capitalism and the Emergence of the Nordic Internal Market[1]

ATLE MIDTTUN

I. Introduction

With the recent liberalisation of Sweden and Finland, in addition to the early market reform in Norway, the Nordic countries (with Denmark gradually attached in a semi-liberal position) now constitute an internal free-trade market. The three Nordic countries are thereby putting into practice a more far-reaching liberal market system which eclipse even the EU Commission's most ambitious plans for an internal European market.

A characteristic feature of the Nordic reforms is their neutrality to ownership. As opposed to the British reform, they did not include privatisation, but largely left public ownership intact. We are in other words dealing with a system dominated by competitive public capitalism, but also with considerable input of private capitalism.

The peculiar mixture of public ownership and competitive exposure reflects a Nordic tradition of pragmatic social democracy. Characteristically, the Nordic reforms were introduced and implemented by both

---

[1] This chapter draws together a broad set of information on three Nordic countries. Detailed referencing is only done for tables and figures, and for special statements. Otherwise the reader will find a list of our Finnish, Norwegian, Swedish and Nordic sources in the Literature list for this chapter.

conservative and social democratic governments, and except for some flagging by the liberal Bildt Government in Sweden, the reforms raised little political controversy. Even in the Swedish case, the reform was implemented by the Social Democratic Carlson Government in the final round.

This chapter gives an overview over the main features of the three liberal Nordic reforms, and describes the basic shaping of the new internal Nordic market. As the Norwegian reform preceded the Finnish and Swedish reforms by five years, and since Norwegian trade institutions have played a major role in the new Nordic market, the Norwegian case is given special attention. After five years of liberal trade in Norway, there are also some experiences to be reported.

II. Structure and Traditions in Nordic Electricity Industries

All three Nordic countries, Finland, Norway and Sweden can be characterised as electricity-intensive countries. With 25,000, 17,000 and 11,500 KWh electricity consumption *per capita*, they rank at the top of the European statistics (Fig. IV.1). In spite of small populations, the Nordic electricity industries therefore make a significant contribution to the north European electricity supply (Table IV.1). One of the reasons for this position is that Norway, Sweden and Finland all have a large heavy industry, based on an abundant supply of cheap electricity. In Norway the dominant electricity-consuming industry has been aluminium, other light metals and fertiliser. In Sweden and Finland, paper and pulp have been the dominant electricity consumers.

The Nordic countries are low-cost producers of electricity (Table IV.2), and therefore expect to be competitive in a European free-trade context. Norway especially, but also Sweden and to some extent Finland, can draw on unique hydro-power resources. Combined heat and power systems based on spill-over from the paper and pulp indus-

Table IV.1. Production and distribution in the three Nordic countries

1994	Sale/distribution (TWh)		
	Norway	Sweden	Finland
Total production	113	153	68
Total distribution	76	63	30
Average net yearly export, last five years	5.68	1.12	−8.04

Source: Fact Sheet and Energy Statistics in Norway, Sweden and Finland (1994).

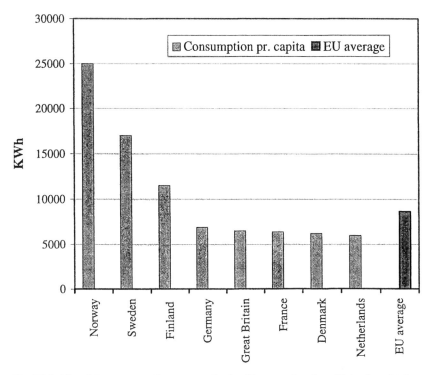

Fig. IV.1. Electricity consumption per capita for Norway, Sweden, Finland and other European countries in 1994. *Source:* OECD statistics (1994).

Table IV.2. Energy carriers and electricity prices, 1994 (excluding taxes, in NOK/KWh) in the Nordic countries

1994	Norway	Sweden		Finland	
Energy carriers	Hydro 100%	Hydro	42%	Hydro	17%
		Nuclear	51%	Nuclear	27%
		District heating	4%	District heating	17%
		Industry	3%	Industry	16%
		Condense	0%	Condense	19%
Household prices	0.48	0.58		0.65	
Industry prices	0.23	0.25		0.38	
Installed capacity (MW)	27,000	34,532		14,000	

Source: OECD (1995), and electricity statistics in Norway, Sweden and Finland.

Table IV.3. Company size and market structure in the Nordic countries

Companies:	Norway	Sweden	Finland
Number of producers	*c.* 129	*c.* 250	*c.* 130
Production (20 largest)	99.3 TWh	148 TWh	60.5 GWh
Average size (20 largest)	4.73 TWh	6.94 TWh	2.74 TWh
Number of distributors	*c.* 200	*c.* 280	*c.* 125
Distribution (20 largest)	42.7 TWh	60 TWh	25.5 TWh
Average size (20 largest)	1.61 TWh	3.10 TWh	1.21 TWh
Market concentration (Herfindahl)	0.32	0.09	0.18

Source: Tema Nord (1994), and electricity statistics in Norway, Sweden and Finland.

try also helps account for low prices in Finland. As far as market structure is concerned, Norway, Sweden and Finland have a large number of producers and distributors (Table IV.3) but they differ considerably in size. On the production side, the Swedish electricity market, given Vattenfall's (the state company) dominant position, can be classified as strongly concentrated[2] (Table IV.3). However, for the Nordic market taken as a whole, the concentration level becomes reasonably low (0.0851 Herfindahl). On the distribution side, the concentration is far lower, and all countries rank low in concentration.

III. The Norwegian Reform: The First Wave of Nordic Liberalisation

III.A. The regulatory model

In June 1990, the Norwegian Storting (Parliament) approved a new electricity market reform. The reform was implemented in January 1991. The basic idea behind the Norwegian reform was to split up different functions of the electricity system according to the consequences of exposing them to competition. Functions which have a natural monopoly character were to be organised as regulated monopolies, while other functions were to be open to competition.

Production and trading were conceived of as activities which could be regulated by competition. The transport functions, transmission and distribution, were seen as natural monopolies, and were hence regulated by administrative control (Fig. IV.2).

For the production function, the securing of efficient competition was

---

[2] By reference to standards set up by US antitrust regulation, market concentration up to 0.1 (Herfindahl's index) is considered low, up to 0.18 medium and above 0.50 high.

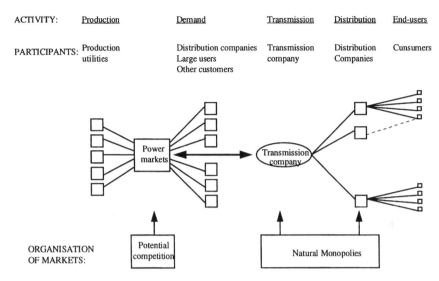

Fig. IV.2. Sketch of the electricity system. *Source:* Hope and Strøm (1992).

therefore seen as the essential regulatory task, the assumption being that effective competition would force the actors to allocate resources productively and to develop efficient organisational behaviour. For the transmission and distribution functions, the essential task was to secure third-party access and to control the efficiency of the natural monopolists.

III.A.a. The Reform in a nutshell
Some essential features of the Norwegian reform were as follows.

1. It included a separation of the high-voltage transmission systems from the State Power Company, Statkraft. The national grid is now owned and administered by a new state company, Statnett, while Statkraft has been reorganised as a pure commercially oriented generating company, although still owned by the Norwegian state.
2. It introduced the principle of common carriage and third-party access for all networks, including regional and local. Suppliers thereby lost their exclusive right to supply their franchise area, and were instead subjected to price competition. Area concessions to grid owners are still combined with an obligation to connect to the national grid and supply power to the area, but the concessions do not allow them to discriminate against deliveries from other suppliers. In return, the generation and distribution companies are no

longer obliged to supply their areas on a long-term stable-price basis.

3. It opened up for consumer participation in the spot market for occasional power which for a long time had operated according to free-trade principles. After the reform, this spot market has been further differentiated, and a set of markets have been established or are now emerging. These markets are now administered by Nordpool, a company jointly owned by Statnett, and the Swedish grid company, Svenska Kraftnät, which acts as a clearing institution.

4. As opposed to the British reform, the Norwegian reform had no restriction on which consumer segments were allowed to trade in the liberalised electricity market. In principle, a household can buy power directly in the spot market for its own consumption, although this has not happened extensively in practice.

5. It forced vertically integrated companies to ring-fence generation from distribution through separate divisions with separate accounts.

III.A.b. What the reform did not change
The basic organisational structure from the 'old' regime still remains, and many of the traditional network ties between the companies are still active. It is, therefore, perhaps just as noteworthy to list what the reform did not change.

1. It did not affect the ownership structure, implying that the main part of the electricity system still remains public.

2. It did not alter concession rules which strongly disfavour transfer from public to private ownership.

3. It did not demand organisational separation of the transmission from the production and trade functions at the regional and municipal levels. This implied a weakening of the control against cross-subsidisation between the monopoly and competitive parts of the electricity industry.

4. It did not increase the staff of the regulatory agency NVE sufficiently to control non-competitive changes effectively.

5. It maintained a foreign trade regime for electricity based on licences for long term agreements, and thereby prevented normal international trade from taking place.

6. It did not open up the special contract market for power to energy-intensive industry.

In sum, the reform did not fully implement a free-trade model, but rather introduced competitive elements into a cost-based, supply-oriented infrastructure. Both public ownership as well as the political and contractual 'embeddedness' of large parts of the electricity trade served to limit the effects of competition. However, the increasing participation of new, commercially oriented traders and brokers, and a strong commercial reorientation of some leading companies is gradually having a general 'commercialising' effect on the electricity industry. A new ruling from the regulatory authorities in 1995, which allowed small customers to buy on so-called predetermined curves, also increased the competitive pressure on the Norwegian electricity system.

It is also worth noting that the Norwegian electricity market reform was part of a larger commercialisation process which also included a strengthening of competition law and competition authorities. A reform of the Norwegian competition authority in January 1994 marked a transition to a more advanced competition-regulation regime.

The competition authorities were also given a central role in supervising trade within the new market regime. Recently, Professor Einar Hope, the intellectual architect of the Norwegian electricity market reform, has become director of the Competition Authority.

III.B. The role of the occasional power market

A general observation on reforms is that they are shaped by pre-reform history. In the Norwegian case, one should particularly highlight the crucial role played by the so-called market for occasional power. The Norwegian electricity system before the reform consisted of three wholesale markets:

1. electricity for general supply, covering approximately 60% of the wholesale supply;
2. electricity for energy-intensive industry, covering approximately 30%; and
3. occasional power, covering 10% of the wholesale supply.

These markets varied from the heavy politicisation of the hydro-power market to energy-intensive industry, to the competitive market for occasional power. The bilateral wholesale market for general supply was organised through municipal and regional monopolies and cost-based trading on long-term contractual relations. In many cases this market was internalised in vertically integrated companies.

The market for supply to heavy industry was organised through long-term contracts between public producers, dominated by the State power company, and private and public energy-intensive industry. The terms of contracts were decided upon through political barter in Parliament.

As opposed to the two other markets, the market for occasional power was organised on an ideal free-trade basis, with an industry-owned organisation, 'Samkjøringen' acting as a clearing house. This market originated from internal trade between producers, and served to optimise the use of their water reservoirs. The exchange prices signalled to each reservoir owner whether to sell or buy. The prices also signalled whether to use electricity or oil for heating large industrial boilers which were usually tied to electricity companies on special flexible contracts.

The Norwegian market for occasional power also had a Nordic extension, as it was also indirectly accessible to Swedish and Danish producers, albeit with the Norwegian State power company as an intermediary. Over 60 Norwegian production companies were trading on this market before the reform.

The crux of the Norwegian reform was that it opened up this market to end-users, thereby supposedly also making it a reference market for the bilateral forward market. (If the actors in the bilateral forward market could move into the spot market, the expected spot price would become a reference price for bilateral trading.) The existence of a fully developed pre-reform spot-market helps explain how Norway managed to develop a reform so quickly.

III.C. The new spot market

One way of presenting the Norwegian reform in practice is to present the outcomes of the reform for the wholesale and end-user markets. We shall briefly describe some of the main outcomes for the spot market, the bilateral forward market, and the end-user markets, to heavy industry and general consumption.

As already mentioned, the spot market for electricity grew out of the existing market for 'occasional' power. After the creation of Statnett, the national grid company, this company also took over the operation of the spot market from the producer co-operative, Samkjøringen. Statnett, the Norwegian grid company and its clearing house subsidiary, Statnett Marked, has until mid-1996 operated three markets for domestic trade.

1. The day market, which was created in January 1992, was a direct continuation of the market for 'occasional' power, which had been in operation since 1971. Actors who wish to trade on this market may register offers and demands for up to one week at a time with the possibility of making corrections on a daily basis. One-day contracts are priced at intervals with varying prices. The price is set

by a market cross based on balance between the aggregated supply and demand curves. The price is adjusted for bottlenecks in the national grid so as to secure physical balance within each area. Since the price is calculated for each geographical area, a capacity charge is automatically integrated in the price setting.

2. The weekly market, which was a continuation of the pre-reform market for 'supplementary' power, was started in the autumn of 1992 in order to open up the export of guaranteed power. This market now includes two types of power: base load, specified to a delivery of 168 hours a week; and day load, delivered from 0700 to 2200 Monday to Friday. Base-load and day-load are treated as two different contracts, with different pricing. Offers and demands can be registered for each week, up to one year ahead. As for the weekly market, the price is set by a market cross based on a balance between the aggregated supply and demand curves. No consideration is taken of bottlenecks in the pricing of weekly power. Since autumn 1995, the weekly market has been turned into a futures market with a daily settlement between the actors and the market maker, Nordpool, according to the market value of the contract.

3. The market for regulatory power is used to co-ordinate the running of the Norwegian power system. In order to achieve satisfactory quality of the power delivered, it is necessary at any moment to have a balance of supply and demand. The actors in the market for regulatory power have to be able to regulate their production given 15 minutes notice. Payment is given by Statnett according to a previously specified offer from the producers, where the cheapest alternative is given priority. The prices on the market for regulatory power have so far been very close to those of the daily market.

As already mentioned, the daily spot market has played an important role in the new regulatory system. Intentionally, it serves both as a market for reservoir management and for direct sales to large consumers and distribution companies. Prices in the spot market contain both a signal to the reservoir managers about the current shadow prices on water, and a signal to the bilateral contract market in terms of expectations to future spot prices as a relevant alternative price to long-term contracting.

The pre-reform market for occasional power was not only organised on a free trade-basis, but actually also performed largely according to free-trade expectations. As can be seen from Fig. IV.3, the market functioned with some sensible relations between prices and supply. In wet periods (with high accumulated production), prices were low and export was high; in dry periods, prices were high and Norway had

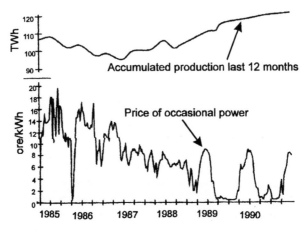

Fig. IV.3. The market for occasional power: prices and production. *Source:* Statnett Marked (1994).

hardly any imports. Prices also clearly followed seasonal patterns, with high winter prices (fourth and first quarters) and low summer prices (second and third quarters), but they were highly volatile. The reform marked no immediate and dramatic change in this picture. A rather dry year in 1991 brought prices up, but to no dramatic peak. The following wet years of 1992 and 1993 brought prices fairly low, triggering low price expectations throughout not only the spot market, but also the market for bilateral forward contracts.

Even though the Norwegian market as a whole is far from perfectly competitive, competition in the wholesale market has been sufficient to force prices in the market system as a whole down to levels in 1991–1993 where some of the generators experienced insufficient revenues to cover their large up-front investment costs.

A sharp price rise during winter 1993–1994 (Fig. IV.4) has set expectations for a higher future price scenario. This price development cannot easily be explained by the hydrological factors, and must probably be seen against the background of new strategic configurations. With the high peak in 1994 as an exception, the spot price has followed a rising trend in 1995 and 1996, both as far as winter peaks and summer bottoms are concerned. Until spring 1996 there seems to be little justification for this price development in the underlying hydrological conditions, as indicated by the accumulated production curve. This may again indicate more clever strategic market behaviour by the producers than in the first post-reform period. However, increasing demand after economic recovery in Norway may also be part of the explanation.

Fig. IV.4. Development in spot prices after the market reform. *Source:* Statnett Marked (1996).

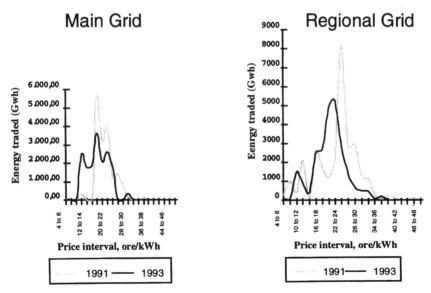

Fig. IV.5. Energy traded by price intervals, for the main and regional grids. *Source:* Midttun and Køber (1995).

III.D. Bilateral contracts in the wholesale market

III.D.a. 1991–1993
The forward market, based on bilateral contracts, has traditionally been the dominant market, and still is. The general picture of price development in the first period since the market reform was characterised by a moderate price decrease and a somewhat increasing price diversity. The price decrease was between 14.04 and 9.35% from 1991 to 1993 (Midttun and Køber, 1995). The price spill-over from the spot market has in other words not been dramatic for the market seen as a whole, although a few companies with highly speculative engagements made considerable losses.

Expectations of a development towards more homogeneous prices as a result of market reform were not fulfilled in the first round. Both in the regional and in the main grid trade, there was a shift from a single price reference for a large part of the trade under the pre-market regime to a spread of contracts over a larger price spectrum after the reform (Fig. IV.5). It seems as if the focus on a political standardised reference price (the selling price of the State power company) under the old regime gave a narrower price spectrum for a large volume of contracts. Traders under the more commercial 1993 regime seem to have lacked a general consensus over such a reference price.

The commercialisation of trade after the 1991 reform did not immediately abolish the large regional price variations which characterised the regional self-supply economy of the pre-reform period (Midttun and Køber, 1995). The large differences in prices in the bilateral forward market indicate that network relations and regional and local institutional structures still, three years after the reform, play a major role in regulating trade. To some extent, price development seems to indicate that commercialisation under varied institutional conditions and under unequal commercial competencies may in fact lead to diversification rather than homogenisation of prices.

III.D.b. 1993–1995
Data for detailed regional split-ups do not exist in the period 1994–1995. However, national data are available for various contractual types from 1993 to 1995 (NVE 1993, 1994, 1995). The picture, in the period 1993–1995 is one of moderately decreasing prices in the first round, and then price increase in the second, reflecting the development which has already been observed in the spot market (Table IV.4).

The price differences between short- and long-term contracts were generally higher in 1994 when the short-term contracts were at their

Table IV.4. Volume and average prices on remaining bilateral contracts

Remaining contractual time when data was collected	1993		1994		Remaining contractual time when data was collected	1995	
	GWh	Average price	GWh	Average price		GWh	Average price
0–2 years	13,532	14.94	26,350	13.01	0–1 years	16,368	16.1
2–5 years	15,148	15.95	6660	15.71	1–3 years	17,195	16.1
5–10 years	7283	16.10	7245	16.50	3–5 years	9253	16.6
10–15	2734	13.63	2178	15.30	5–10 years	8006	17.3
15+ years	4153	15.71	2158	15.70	10+ years	3673	17.4
Sum	42,850		46,791		Sum	54,495	
Weighed average		15.49		13.39			16.5

Source: Electricity Market Survey for 1993, 1994 and 1995 (NVE 1993,1994 and 1995).

lowest, indicating that actors have upheld medium to high long-term expectations in spite of low short-term prices.

III.E. Results for end users

In the final round, the reform will tend to be judged by its consequences for end users. Here we have to distinguish between at least two end-user segments: on the one hand energy-intensive industry, and on the other hand light industry/services and households.

III.E.a. Energy-intensive industry

Norway has traditionally treated its abundant hydro-resources as a competitive advantage to be harvested through energy-intensive metal production. In the pre-war period, foreign companies were heavy investors in Norwegian aluminium production. Due to patriotic sentiments after Norwegian independence from Sweden in 1905 and nationalistic attitudes towards the use of Norwegian resources, the so-called concession laws were passed in 1917, limiting private industrial access to hydro-power. Norwegian electricity-intensive industry thus came to rely heavily on electricity generated by the State power works.

The post-war industrial policy saw energy-intensive industry as the locomotive of the Norwegian economy, and an extensive public hydro-power production program was established to provide the necessary electricity input. This electricity was generally provided on a long-term contractual basis at very favourable terms. Since the 1950s these contracts have been standardised and approved by Stortinget (Parliament). In the 1970s the paper and pulp industry also received access to this 'market'.

Table IV.5. Renegotiation of electricity prices and savings thereby for households and light industry

	1993	1994
Percentage of households that have renegotiated electricity-prices or changed supplier	1.1	6
Savings achieved by renegotiating (%)	0	7
Savings achieved by changing supplier (%)	0	2
Percentage of enterprises which have renegotiated electricity-prices or changed supplier	42.0	49.3
Savings achieved by renegotiating (%)	18	1
Savings achieved by changing supplier (%)	25	8

Source: NVE (1994).

As already mentioned, the Government explicitly exempted this heavy industry 'market' from the reform, and the practice of giving energy intensive industry politically guaranteed prices through special contracts continued. Stortinget (Parliament) thus approved of a new round of contracts in 1992, running to 2010. For about one third of the power produced in Norway, the market reform therefore had little or no impact. The reform did, however, have an impact on heavy industry through its consequences for spot-market prices, since this industry traditionally buys some of its electricity there.

III.E.b. Households and light industry

Households and light industry were in principle more directly affected by the reform. Both segments were formally given a new right to participate in the spot market either directly or indirectly through brokers or any regular electricity company, including their local grid-owner. But in practice, metering costs and information/competence costs largely kept households away. Whereas large sections of light industry, including services, renegotiated contracts and obtained substantially lower prices, households lagged far behind (Table IV.5).

Industry took out its price savings early (1993) and gained little in the next round in 1994. The few households which managed came late and gained nothing in 1993, but more in 1994. Notably, households gained more from renegotiating than from changing supplier. Household participation in the spot market probably reflects the fact that some companies offered the households so-called market-based prices. However, large and concentrated housing complexes may have renegotiated collectively through brokers.

Price development for both light industry and households since 1993 has first been stable and has then risen (Table IV.6), indicating that the

Table IV.6. Prices to households and light industry in Norway in øre/kwh.

	1993	1994	1995	1996
Power price (households)	15.2	15.4	17.9	17.8
Power price (manufacturing services)	13.6	12.6	15.7	—
Transmission cost (households)	20.2	18.6	18.2	17.6
Transmission cost (small manufacturing and services)	17.7	16	16	14.8
Transmission cost (large manufacturing and services)	12.5	11.6	11.6	11.1

Source: Electricity Market Survey for 1993, 1994 and 1995 (NVE 1993,1994 and 1995).

Table IV.7. Total prices to households

	1990 (old regime)	1991	1992	1993	1994	1995	1996
Household prices	35.1	35.6	35.3	35.4	34.0	36.1	35.4

Source: NVE (1996).

price increases in the wholesale markets are gradually spilling over to end-users.

The households only achieved 'real' market participation after a Directive from the regulatory authority, NVE, in 1995, which allowed them to buy on so-called pre-determined curves. They therefore entered the market too late to profit substantively from the low price market. The manufacturing and service industries, which had initially profited from the market reform, were eventually forced back to a higher-price scenario. Transmission costs have come down a little (Table IV.6), and it is generally believed that they have a future efficiency potential to be harvested through possible new price-cap oriented regulatory reform.

All in all, household consumers have had little to gain from the market reform, and prices have remained more or less stable in spite of the transition to the new regime (Table IV.7).

IV. The Finnish and Swedish Reforms: The Second Wave of Nordic Liberalisation

Finland (in spring 1995) and Sweden (in January 1996) implemented electricity market reforms very much inspired by the Norwegian initiative, but also modelled on the British and the liberal EU initiatives. As in Norway, the electricity market reforms in both countries replaced planned-economy systems with decentralised regional monopolies, supplemented by state-company engagement, especially on the production side.

In Sweden the old electricity law, dating back to 1902, had given the regional and local grid-owners monopolistic concessions to serve customers in their areas. This monopoly was balanced by an obligation to deliver to all customers at reasonable prices. The responsiveness of the system to societal interests has partly been guaranteed by public ownership, and partly by a formal right to complain to state price and regulation authorities (Fritz and Springfeldt, 1995).

Under the new regime, Närings och Teknikutvecklingsverket (NUTEK) regulates the electricity system and allocates grid licenses to local and regional companies. Trade within the Swedish system is in principle free, but export contracts with a duration over six months must be notified to Svenska Kraftnät, the national grid company.

In Finland, the market reform replaces a planned-economy based electricity regime both at the national and regional level, introduced by a law dating from 1979/80. Before 1979, production and distribution of electricity had not been legally regulated except for safety aspects (the law of 1928). The planned-economy system was based on licensing for electricity production and for grids. The country was divided into 20 regions, each with a regional commission for electricity supply, where distributors, wholesalers and producers within the region were represented.

National producers and developers of the high-voltage grid reported directly to the Ministry of Trade and Industry. The Ministry also acted as a co-ordinator of regional plans. Electricity companies operated under a concession from the Ministry which obliged it to supply its distribution area. Import and export rights were given by Government, and only the two large central grid companies Imatran Voinam Oy (the State company) and Teollisuuden Voimansiirto Oy (a company controlled by heavy industry) had import rights, except for some minor import from electricity distributors in the border area for local purposes.

IV.A. Characteristics of the Finnish and Swedish reforms

Some of the essential features of the Finnish and Swedish reforms were as follows.

1. The Swedish reform included a separation of the high-voltage transmission system from the State power company, Vattenfall. The national grid is now administered by a new State company, Svenska Kraftnät. As opposed to its Norwegian sister company, Svenska Kraftnät does not operate the high-voltage grid, but relies heavily on operators working on a contractual basis, especially Vattenfall.
2. Finland is in the process of reorganising the public and private high-voltage grid companies IVS and TVS into a common company.

As opposed to the Norwegian and Swedish solutions, the Finnish company will have a mixed private and public ownership. The State power company, Imantran Voima Oy (IVO) and the industry will have a third of the shares each, while the State's direct involvement and institutional investors will account for the remaining part. The new grid company will thereby be an independent company with the system responsibility transferred to a new institution, the Electricity Market Centre.

3. The Swedish and Finnish reforms, like the Norwegian, opened up common carriage and third-party access for all domestic networks (national, regional and local). In Finland the low-voltage grid (below 110 kV) is regulated by the Electricity Market Centre (EMC), which grants operating permits to the local transmission companies within specific areas. In Sweden, the low-voltage grid (below 130 kV) is regulated by NUTEK.

4. The Swedish reform demanded full organisational separation of the grid on the one hand and sales and production services on the other. The Finnish reform here followed the more lenient Norwegian reform in demanding separation by accounting only.

5. The Swedish and Finnish reforms were more moderate than those in Norway as they did not allow full participation from small household customers at the outset. In Sweden, monopolised delivery concessions to small customers were granted to the local utilities for a period of three years. In Finland small customers (below 0.5 MW) were exempted from free trade during the first one to two years. The Swedish and Finnish reformers here obviously chose to follow the gradualistic British approach.

As opposed to Norway, Swedish and Finnish liberal reformers could not fall back on an already existing spot-market trade and spot-market institutions. This may be one of the factors which accounts for their slower implementation of liberal trade. In both countries, a carefully negotiated merit-order system had to be replaced. Sweden and Finland here took different paths. The Swedish national grid company quickly entered into close co-operation with its Norwegian sister company, and used the Norwegian Bourse as a stepping stone towards a common Nordic system. Finland, perhaps because it lacked a strong independent national grid system which could take on Nordic ambitions without considering national producer interests, chose to develop its own Bourse.

After the Swedish and Finnish reform, the pricing of grid services in the whole Nordic system is based on point tariffs, implying that customers face one tariff for each grid level. This system is practised in its pure form in Norway, and in a somewhat modified form in Sweden, where it is coupled to a geographical differentiation, implying that producers face relatively high tariffs for electricity input in the north of Sweden and low tariffs in the south. Finland has adopted the Norwegian model of a point-tariff system without geographical differentiation. The Government and smaller electricity suppliers have been in conflict with TVS and IVS, who wanted distance to be taken into account in the point-tariff clause.

The reforms have left system responsibilities with the grid companies (in Norway with Statnett, and in Sweden with Svenska Kraftnät), but the short-term balancing is done according to somewhat different principles. In Norway, Statnett handles this balancing through an auctioning process in the so-called market for regulating power. In Sweden, Svenska Kraftnät handles the balancing service through instructions to specially designed producers based on contractual agreement. The responsibility for the short-term power balance in Finland lies with IVO. Due to IVO's dominant position there is no market for regulating power as in Norway. System responsibility in Finland is, as in Norway and Sweden, left with the national grid companies. However, with two parallel grids, system responsibility in Finland remains with two companies. After the merger between the two grid companies, IVS and TVS, system responsibility will be transferred to the Electricity Market Centre.

As far as regulation is concerned, Sweden has followed the Norwegian model of delegating regulation to an extra-ministerial authority, NUTEK. However, as opposed to the Norwegian authority, NVE, NUTEK enjoys cross-sectoral competence and also supervises other industrial sectors. Regulation in Finland, after the market reform, lies with the Electricity Market Centre. In spite of its limited size, the Centre has a broad mandate to supervise and regulate the new market regime. The Centre covers its costs through service fees from the power sector. By assigning the regulatory follow-up of the electricity market to a specialised agency, Finland has followed the Norwegian rather than the Swedish model.

IV.A.a. Unaffected ownership relations

As in Norway, the Swedish and Finnish reforms did not affect ownership. This implies that the main part of the electricity systems both in Finland and Sweden still remains public, although with a larger share of private ownership than in Norway. However, Finnish and Swedish ownership has not traditionally been protected by concession rules which, as in Norway, strongly disfavour transfer from public to private ownership. We may therefore expect to see a more dynamic development of ownership in the Finnish and Swedish cases as a consequence of the new market economy.

IV.A.b. New competition law

As in Norway, the new Finnish and Swedish electricity regimes were part of a commercialisation process which also included the strengthening of competition law and competition authorities. In 1988, Finland reformed both the legal basis and the organisation of its competition authorities. The goal was to support 'healthy and functioning economic competition'. The competition authorities have been involved in electricity issues, both through intervention in pricing and terms of agreement in electricity trade. Of particular importance is its intervention in the issue of load management and system optimisation in 1993, and in the terms of wholesale trade imposed by the State company, Imantran Voima.

Sweden reformed its competition law in 1993, following the general competition rules of the European Union. Two core principles were implemented: firstly, a general prohibition of collaboration intended to undermine competition, and secondly, a prohibition of the misuse of a dominant market position (Fritz and Springfeldt, 1995). The relatively new application of competition rules to the electricity market implies that there were not enough decisions to establish a clear conceptualisation of Swedish competition practice in the electricity field.

IV.B. The background and politics of Nordic reform[3]

Politically, the Nordic reforms were characterised by pragmatic adaptation rather than ideological debate. The Conservative Swedish Bildt Government's use of the electricity-market reform to profile its liberal policy was a passing exception. Important reform elements were initiated by the Social Democrats, who also undertook the implementation of reform. The fact that all three Nordic reforms left ownership relations and the organisational structure of the electricity industry intact helps to explain the broad political consensus.

On the professional level, the reforms were manifestations of an economic paradigm change from a macro-oriented planned-economy model to a more micro-oriented self-regulatory regime. They also redefined the electricity industry as primarily economic rather than primarily technical issue.

---

[3] The Norwegian and Swedish parts of this Section are largely built on a more extensive presentation in Midttun (1996), which contains extensive interviews with central reform actors. The Finnish part is extensively based on Gundersen (1996).

IV.B.a. The Norwegian reform background

The Norwegian reform, in its first phase, was very much administratively driven. Central officials within the Ministry of Finance had for some time been dissatisfied with investments and pricing decisions within the existing electricity regime. Inspired by the British reform and EEC initiatives to liberalise European electricity markets, the Ministry of Finance, in collaboration with the Ministry of Industry and Energy, in 1988 ordered an investigation of a possible Norwegian market reform from the Norwegian School of Economics and Business Administration.

The investigation ended up with a recommendation for a decentralised competitive electricity market which was later adopted as part of the reform. The new market orientation implied a complete turnaround by the central authorities' who had previously advocated centralised reform aiming at hierarchical integration of the decentralised Norwegian electricity supply into 20 vertically integrated companies.

A law with some market elements included was proposed by the Social Democratic Brundtland Government in September 1989. The Conservative/Centre Party/Christian Democratic Syse government (with Eivind Reiten, a former director of the energy division of Norsk Hydro as Energy Minister) thought the reform was too limited and withdrew the energy proposal. In March 1990, a new and more radical market reform was presented by the Syse Government, and approved in Parliament in June of the same year with little opposition. The law was implemented under the new Brundtland Government without any changes.

One of the political bases for support for the Norwegian reform has clearly been the Norwegian municipalities and their electricity companies. As already mentioned, the market reform replaced plans for a dramatic centralisation of Norwegian electricity industry. The proposed centralisation policy ran contrary to the traditional structure of the electricity industry, where Norway's approximately 320 electricity companies had traditionally pursued independent local and regional policies.

The other base for support came from industry, and was clearly expressed by its interest organisation. The reform, which allowed industrial customers to opt out of local monopolies and to operate freely on wholesale markets, effectively curtailed the traditional political practice of price discrimination against small and medium-sized industry to the benefit of households. The industrial lobby also expected to benefit from lower prices due to power surplus. Finally, many energy producers saw the market reform as an opportunity to increase their commercial freedom and their independence from political ties.

IV.B.b. The Swedish reform background

Like the Norwegian reform, Sweden's was also solidly anchored within the public administration. During the 1990s, officials in the State Energy Board took the initiative to define and describe a market-oriented regime for electricity. Elements of a market-based reorientation of the Swedish electricity system were also present as part of a social democratic policy for economic growth in 1991.

However, the reform process gained further political momentum with the coming to power of the Conservative coalition Government in autumn 1991, and the coincidence between the professional and political paradigm changes had an amplifying effect which led to the presentation of a White Paper on 'An Electricity Market with Competition' (Ministry of Industry, 1991–92b). The goal was, through competitive reform of the Swedish electricity market, 'to reach a more rational exploitation of the production and distribution of resources and to secure customers flexible delivery conditions as the lowest possible prices'. This policy won wide support in Parliament, which, however, expressed concern about reciprocity in international trade under liberal governance, and was critical of private ownership of the central grid company.

In spring 1994, after a proposition from the sitting Conservative–Centre coalition Government, Parliament voted in favour of a new liberal electricity market regime, to be implemented by January 1995. However, after the 1994 election, which led to a Social Democratic Government, implementation of the market reform was postponed in order to await further consequence analyses. A proposition on the financing of the regulatory agency for the new electricity market was also withdrawn. The consequence analysis was undertaken by the Energy Commission, which presented its results in early 1995 under the title 'New Electricity Market'. The Commission recommended that the resting electricity law be implemented with only minor changes. The Social Democratic Government thereafter presented a new law proposition along the guidelines of the old one in May of the same year, which was supported by Parliament in the autumn of 1995. The reform was put into practice in January 1996, with the understanding that adjustments and expansion of consumer access should subsequently be undertaken.

As in Norway, Swedish industry actively lobbied for electricity market reform, with the expectation of lower prices in a surplus-ridden production system with low, variable unit costs. Publicly controlled electricity producers and the State-owned Vattenfall in particular, could, like its Norwegian equivalents, hope to gain a freer position *vis-à-vis* its political owners. However, the large Swedish producers wished to maintain their traditional collaboration over production optimisation, and objected to extending the market reform to the wholesale trade among producers. On this point, however, they were efficiently over-ruled and Government, on Svenska Kraftnät's recommendation, adopted a 'Norwegian' market-based solution.

IV.B.c. The Finnish reform background

Reform discussions in Finland date back to the early 1990s. In April 1992, the Government signalled the necessity of an electricity-law revision, and the Electricity Commission in its statement of the same spring suggested a reform. Government made a definite commitment to reform in its energy policy statement to Parliament in 1993, and the new reform proposal was presented in autumn 1994 . The law was passed in the spring of 1995, and has been operative since mid-1995.

As in Norway, the introduction and implementation of the Finnish Electricity Market Act raised little political controversy. The reform was proposed by the Social Democratic Government in the early 1990s, but with the change in Government, the new Centre Party Government (led by President Escko Aho) in coalition with the conservative National Coalition Party, put forward the final proposal. The law was agreed on by the Parliament without significant dispute early in 1995. The only critical questions raised in Parliament were about the security of the smaller consumers and how the reform would affect the electricity supply in the districts. However, this debate did not threaten the reform, but slightly altered its content.

Although the reform raised little controversy in party-political terms, there were disputes between electricity consumers and the production industry over its implementation. Electricity companies lobbied for a delay of one to two years in order to permit time for present supply contracts to expire and readjustments to be made. Lobbying efforts by the electricity utilities produced their first concrete result in October 1993, when the Ministry of Trade and Industry (MTI) agreed to delay the new legislation until the whole issue was examined in 'closer detail'.

On the other side, SYKL, the central association for small to medium-sized companies, accused Finland's electricity supply board of attempting to derail the Government's plan to introduce the Electricity Act. SYKL, which had the most to gain from a quick deregulation, intensified its lobbying to have the MTI reverse its decision to defer the issue, and wanted a quicker implementation of the reform. However, the electricity utilities with their public ownership could raise stronger political support, and won through with their demand for a deferral of implementation.

V. The Internal Nordic Market

The existence of three neighbouring liberalised systems with extensive interconnections, inherited from the old NORDEL co-operation,[4] has created the preconditions for an integrated Nordic electricity market. Since the advent of the Nordic market reforms, this opportunity is being actively exploited by the build-up of Nordic trade institutions and a flourishing inter-Nordic trade. In the first round, the Norwegian and Swedish markets were merged through a common Bourse arrangement, owned and jointly administered by the two national grid companies, Statnett and Svenska Kraftnät. The Finnish system is still not fully integrated with the Swedish–Norwegian system, due to limited and monopolised grid interconnection, but is in the process of expanding its links.

The extended freedom of commercial trade between the three countries is creating structural and strategic challenges to their electricity industry. In terms of size, ownership structure, industrial configuration and restrictions on ownership transfer, the three countries differ significantly, and their industry therefore participates in inter-nordic competition on somewhat different terms.

V.A. The institutions and structure of Nordic trade

V.A.a. The Nordic Bourse
In April 1996, the Swedish national grid company, Svenska Statnät, bought 50% of the shares of the Norwegian Bourse, which was thereby transformed into a Nordic Bourse under the name Norpool under joint Norwegian–Swedish administration. The Bourse undertakes the administration of organised Nordic markets. It secures objective pricing and equal treatment of all players, risk control and monitoring, as well as guaranteeing settlement and supply.

The Bourse per medio 1996 registers 97 Norwegian, 22 Swedish and one Finnish player, including both producers, distributors, industrial companies, market makers, brokers and clearing customers (Table IV.8). The Bourse is becoming a central part of the Nordic market and the volume traded on the Bourse has increased substantially, from a total volume traded of about 17 TWh in 1993 to above 40 TWh in 1995, with a stable increase in both spot and futures trade.

Both the number and variety of actors from the three countries indicate the maturity and experience of the Norwegian actors compared to the Swedish in dealing with electricity trade through organised markets.

---

[4] NORDEL is a Nordic parallel to the west European UCPTE system.

Table IV.8. Actors on the Nordic Bourse

	Norwegian	Swedish	Finnish	Total
Producers	34	8	1	43
Distributors	36	7	—	43
Industrial companies	12	2	—	14
Market makers	3	—	—	3
Brokers	12	4	—	16
Clearing customers	—	1	—	—
Total	97	22	1	120

Source: Statnett, Lecture by Trade Director Hans Randen at the Norwegian School of Management, (April 1996).

The exclusive participation of Imantran Voima in Finland illustrates the *de facto* continuation of considerable State control of foreign trade with electricity, in part through control of grid connections to other Nordic countries.

V.A.b. The Finnish Bourse
Finland has chosen not to join the Swedish–Norwegian Norpool, but to develop a national Bourse—Electricity Exchange Oy (EL-EX). EL-EX is owned by the Finnish Bourse and clearing company SOM, with 80% Finnish ownership share and 20% owned by Optionsmeklarene. A large number of producers, distributors and industrial companies have signed up as members. The new Bourse will allow seasonal, weekly and hourly trade based on forward contracts. System regulation will be secured by an independent balancing company. The Finnish Bourse is technically ready to start up in June, but has to await a legal amendment to allow the actors to trade directly on the Bourse. EL-EX has expressed a wish for some kind of association or collaboration with the Nordic electricity Bourse, which is also in favour of such a collaboration, but it not yet clear if and how this will happen.

V.A.c. The bilateral market
The Nordic electricity reform differs from the British in so far as the actors have been allowed to trade outside of publicly organised markets. This trade has by far outweighed the Bourse trade, even for the more commercially developed Norwegian market. Bilateral trade allows greater flexibility of contracts than the organised markets, which operate with standardised contracts in bilateral trade. Actors may thus design load-management regimes over short- or long-term horizons ranging up to ten years or more, where they can tailor contracts to the needs

and interests of the two actors involved. The actors which participate in bilateral trading are largely the same as those participating in the Bourse trade. Many actors thus participate in both markets at the same time.

V.A.d. The structure of inter-Nordic trade

Since 1963, The NORDEL co-operation has allowed Denmark, Finland, Norway and Sweden to organise extensive electricity trade within the Nordic region. This trade has traditionally consisted of marginal exchange of surplus to balance the independent national production systems, based on short-term marginal costs as exchange prices. The typical trade pattern has been for Norway to figure as a considerable net exporter to Sweden and Denmark, and Denmark and Finland as net importers. Sweden has largely matched its sales to Denmark and Finland through the import of Norwegian power (Fig. IV.6). It is also worth noting that in terms of volume, Norwegian–Swedish trade has been dominant, with Danish–Swedish and Finnish–Swedish trade in second and third positions, respectively. Danish–Norwegian trade, which is the most recent, has been the least extensive.

The basic structure of inter-Nordic trade does not seem to have been affected by the one-sided liberalisation of the Norwegian electricity market in 1991. As far as the new integrated liberal market since the Swedish reform in 1996 is concerned, it is too early to pass judgement. The number of transactions between Norwegian and Swedish actors is obviously increasing, but this need not affect the basic electricity flow.

V.B. Actors in the Nordic trade

V.B.a. Size, ownership and historical legacy

We have seen (in Tables IV.1–IV.3) that the Nordic electricity industries differ considerably in terms of size. The 20 largest Swedish producers are on average more than 1.5 times the size of the Norwegian, which again are on average more than double the size of the Finnish companies.

This implies that Finnish companies are thinly represented among the large Nordic producers (Fig. IV.7). Swedish companies play a dominant role in the top and upper medium stratum, whereas Norwegian companies take a dominant position in the medium and lower stratum. The large Swedish distributors also outsized their Norwegian and Finnish competitors by a factor of more than two to one, and a similar picture could be drawn of the major Nordic distributors.

Considerable differences also characterise ownership of the Nordic

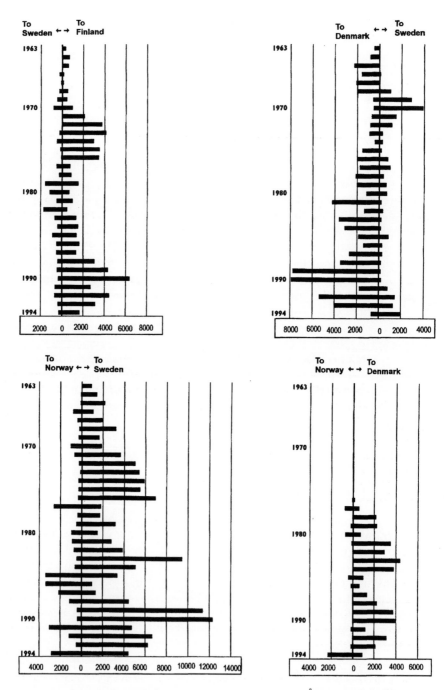

Fig. IV.6. Power trade in NORDEL, 1963–1994. *Source:* Nordel Årsberetning (1994).

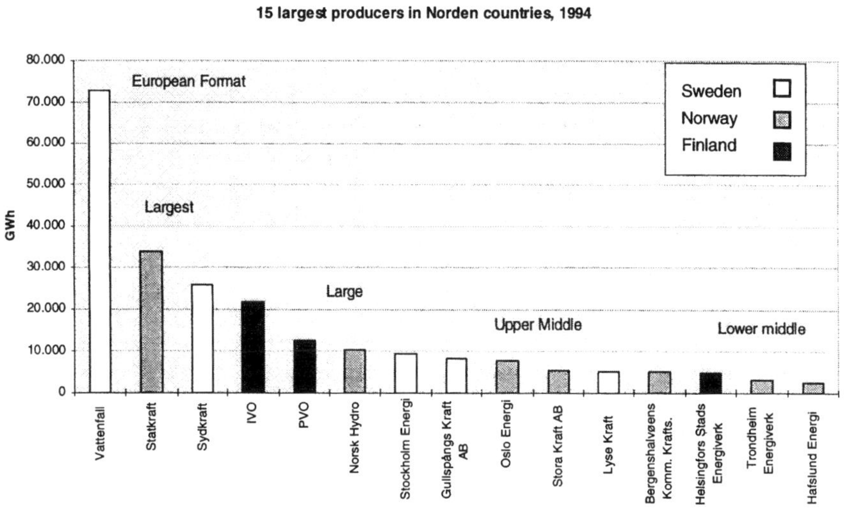

15 largest producers in Norden countries, 1994

Fig. IV.7. The 15 largest Nordic producers. *Source:* Annual reports from the 20 largest producers for 1994.

Table IV.9. Ownership in Nordic electricity production, 1994

	Ownership structure for Nordic producers					
	Institutional (%)	State (%)	Municipal/ county (%)	Electricty sector (%)	Industrial (%)	Unknown
Norway	0	41	40	7	12	0
Sweden	13	55	17	6	6	3
Finland	2	44	15	10	27	1

Source: Annual reports of the 60 largest Nordic producers (1994).

electricity industry, but they are also balanced by strong similarities (Tables IV.9 and IV.10).[5] The State plays a dominant role in Finnish, Norwegian and Swedish electricity producers. However, the Swedish State company, Vattenfall, with its ownership of 55% of the electricity

---

[5] Because it has not been possible within the resources available to analyse all Nordic electricity companies, this section is based on a sample of the 20 largest producers and distributors in each of the three Nordic countries. Because of the high level of concentration, this sample constitutes, respectively, 97% and 36% of Swedish producers and distributors. In the less concentrated Norwegian and Finnish markets the sample constitutes 80% and 74% of the toal volume for producers and 43% and 38% for distributors for Norway and Finland, respectively.

Table IV.10. Ownership in Nordic electricity distribution, 1994

	Ownership structure for Nordic producers					
	Institutional (%)	State (%)	Municipal/ county (%)	Electricty sector (%)	Industrial (%)	Unknown
Norway	0	0	100	0	0	0
Sweden	24	7	45	18	2	4
Finland	2	0	81	12	6	0

Source: Annual reports of the 20 largest distributors (1994).

production among Sweden's 20 largest producers, is in a far stronger position than its Finnish and Norwegian colleagues Imantran Voima and Statkraft, which control only a little more than 40% of the electricity production among the 20 largest companies, and an even smaller share of national production.

The 20 largest Norwegian electricity producers stand out with a strong municipal ownership (40%), a distinctive result of the strong local initiative in Norwegian electricity production. Municipal owner-ship plays a far more moderate role among the 20 largest Swedish and Finnish producers (17% and 15%, respectively).

Institutional ownership, through banks, pension funds, etc. plays a considerable role in Swedish electricity production (11%). This consti-tutes a potential for a capitalist component in the strategic shaping of Swedish electricity industry which is more or less unparalleled in Norway and Finland.

Finnish electricity production is particularly characterised by heavy industrial ownership (27%) as a result of the close ties between the paper and pulp and electricity industries. Such ties are also present in Norway, but to a far lesser extent (12%). In Sweden, industrial owner-ship has been declining, and is now only about 5%. Ownership ties between electricity producers are also present in all three countries, but play a more central role in Finland (10%).

On the distribution side, Norwegian companies are completely municipally or regionally owned. Municipal/regional ownership is also dominant among the 20 largest companies in Finland (81%), but plays a less prominent role in Sweden (45%). Swedish electricity distribution is characterised by extensive State ownership. Vattenfall is not only the largest Nordic producer, but also the largest Nordic distributor. Institutional ownership is a distinct characteristic of Swedish electricity distribution, as it is of production. Swedish distributors are furthermore characterised by considerable cross-ownership within the electricity industry. The characteristic Finnish industrial ownership is also carried over from production to distribution, although it plays a less prominent role.

The differences in ownership of Nordic electricity companies reflect deeply rooted historical traditions. The historian Lars Thue (1995) emphasises differences in the character of natural resources and their location as a factor of central significance for the organisational patterns of Nordic electricity industry. In the early phase of electricity construction, hydro-power constituted the central resource base for all countries. Compared to Sweden and Finland, the Norwegian resources have been better spread out geographically relative to the localisation of the population, more varied with respect to size, and easier to exploit. This has provided a better basis for small-scale electricity production alongside larger projects, and thereby a better basis for local governance of the electricity supply in Norway than in the two other Nordic countries. The less attractive Swedish and Finnish hydro-power resources favoured institutions and organisational solutions with more input of capital resources and expertise.

Thue (1995) also emphasises political and legal conditions as central background factors when trying to understand dissimilarities in the development of the three Nordic electricity industries. In Norway, local democracy has traditionally enjoyed greater economic independence than in Sweden and Finland, where local political initiatives for a long period suffered hindrances. Right up to after World War II, Swedish country municipalities were not allowed to support electrification, because this violated the principle of equal treatment of the inhabitants. Similar rules were also applied in Finland. The Norwegian decentralised municipal model has also been strengthened by the strict concession rules imposed against foreign investment at the beginning of this century. This legislation provided a basis for dominant public influence in hydro-power construction.

The development of the nuclear industry in Sweden and Finland over the past two decades has probably also contributed to the establishment of different organisational patterns in the Nordic electricity industries. Nuclear investments have implied a greater need for capital and competence, and have necessitated regional, national and even Nordic co-ordination. Norwegian industry has not taken the same 'quantum leap', and has therefore been far more able to continue the combination of small and large projects, where small and medium-sized projects have been undertaken through co-operative organisation, by municipalities and regional actors, whereas the State has generally been involved in developing large power projects.

V.C. Strategic developments

V.C.a. Market evolution from traditional network ties
In all three Nordic countries, the transformation to a commercial market economy has been considerably softened by networks including contractually integrated, ownership integrated, and vertically integrated producer–distributor relations (Midttun and Summerton, 1995; Gundersen, 1996). Instead of opting out of old networks, therefore, many Norwegian distributors chose to stay loyal to old producer ties, and adapted to new market conditions through renegotiations, such as most distributors in the Akershus region *vis-à-vis* their supplier, the Akershus Energy Company. Price agreements in many cases deviated considerably from market prices, but were gradually driven in the market-price direction. In a study of adaptation to the market reform in four Norwegian counties (Midttun, 1995) the motivation for such network-loyal strategies among distributors seemed to lie in the perceived lack of competence to operate directly on the wholesale markets, in the perceived risks of such operations.

In Sweden, Summerton (1994) and Midttun and Summerton (1995) found that instead of seeking a competitive strategy by, for example, negotiating with alternative power suppliers, several distribution companies (for instance Gothenburg Energy) chose to strengthen their traditional network of contractual ties with their traditional suppliers among the big Swedish power producers.

Finland followed much the same path as its Nordic neighbours by renegotiating contracts with its previous producers. According to a study made by the Ministry of Trade and Industry (1996), only 6% of those companies who had a consumption above 500 kW and who were able to choose, had shifted supplier. For instance, most of IVO's distribution customers have renegotiated their contracts with the company, but at a lower price than previously. As in the Norwegian and Swedish cases, commercial market forces have therefore to a large extent been mediated through networks and vertically integrated trade.

In all three Nordic countries, however, the traditional network relations have been supplemented and transformed to fit new commercial opportunities. In some cases old network relations have also been deliberately abandoned to the advantage of aggressive strategic behaviour. The overall picture of strategic development is therefore rather mixed, reflecting the interplay between old pre-market and new post-market forces, and the interplay between actors with considerable structural and institutional differences. In the following, we shall briefly present some of the new developments in each of the three countries.

V.C.b. New Norwegian patterns

In Norway, extensive collaboration among producers has emerged as part of export strategies to continental producers with monopoly positions within their domestic markets. Due to their export orientation, this producer collaboration has been accepted by the competition authorities. Currently, there exist four large organisations of generators. According to the Price Directorate, these control more than 70% of yearly average production. If we exclude electricity for power-intensive industry, production is concentrated as described in Fig. IV.8.

Mergers between generators have not been a widespread strategy. There have, however, been a few acquisitions where municipalities have sold their electricity companies to a generator/wholesale company, and there have been a few local mergers between distributors. The slow process of merging between generators is probably due to resistance from the political owners who dominate the Norwegian electricity supply. Attempts have been made to transform two of these groups, SørKraft and OTB-Kraft, into strategic alliances with a focus on the domestic market. However, the competition authorities have, until recently, objected to this. In 1996, however, they reversed the decision and allowed one of the groupings, SørKraft, in response to the opening of a common Norwegian/Swedish market.

V.C.c. New Swedish patterns

Both Sweden and Finland have seen a more gradual adaptation to market competition than Norway, because of the drawn-out period of

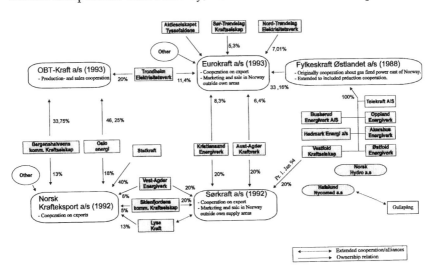

Fig. IV.8. Market share for major producer groups in Norway.

decision-making and implementation. For large Swedish producer-distributors, the long time period has provided an opportunity to strengthen resources and expand markets in anticipation of the market reform (Midttun and Summerton, 1985). Many large power producer-distributors have made extensive strategic purchases of municipal companies in order to expand their service areas. Producers have also engaged in corporate expansion to foreign markets through alliances with foreign producers, shares in foreign power plants, and acquisition of foreign distribution companies. To strengthen their positions *vis-à-vis* producers, many Swedish municipal distributors have formed alliances (joint companies, and joint power-purchase negotiations) with other distributors, resulting in a large number of informal or formal regional 'clusters' of co-operating companies (Midttun and Summerton, 1996).

The most significant change in Swedish production sector was the 1993 merger between the second largest producer, Sydkraft AB, and the fifth largest producer, Båkab AB (Fig. IV.9). Through the acquisition of Båkab from SCA, Sydkraft increased its hydro-electric power resources by 40% to 11.2 TWh. Before the merger, Sydkraft had 16 TWh of nuclear power at its disposal. The Båkab deal also included 5% of the shares in OKG, which meant that Sydkraft's stake in OKG exceeded 50%. In distribution, Sydkraft acquired Malmö Energi, a distributor ranked as the fifth largest in Sweden. This purchase further consolidated Sydkraft as the second largest distributor in Sweden after Vattenfall, with a distribution of above 7000 GWh.

Another significant merger took place in 1992 between the sixth largest producer, Gullspång Kraft, and the seventh largest, Uddeholm. The new company, Gullspång Kraft, doubled its production capacity to 8300 GWh and became the fourth largest producer in Sweden (Fig. IV.10). In distribution, Gullspång has transferred from being an insig-

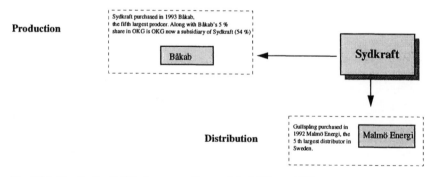

Fig. IV.9. The Sydkraft–Båkab merger. *Source:* Atle Midttun, 1996.

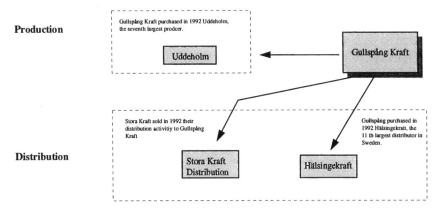

Production

Gullspång Kraft purchased in 1992 Uddeholm, the seventh largest prodcer.

Uddeholm

Gullspång Kraft

Distribution

Stora Kraft sold in 1992 their distribution activitiy to Gullspång Kraft

Gullspång purchased in 1992 Hälsingekratt, the 11 th largest distributor in Sweden.

Stora Kraft Distribution

Hälsingekraft

Fig. IV.10. The Gulldpång–Uddeholm merger. *Source:* Atle Midttun, 1996.

nificant distributor towards becoming the fourth largest distribution actor in the country. Gullspång's new role as a large distributor came as a result of the purchase of Hälsingekraft and the takeover of the distribution division of Stora Kraft.

In addition to the rearrangements discussed above, there have been a great number of other mergers, acquisitions and strategic alliances with importance for industrial structure. Vattenfall has been active both on the international arena and downstream domestically. Stockholm Energi, the third largest producer and distributor, is active domestically in both upstream and downstream activities.

There are tendencies towards a consolidation between production and distribution companies within the same municipality. Examples of this are Västerås and Norrköping. Another strategic pattern is that companies with core business activities within areas other than electricity sell this off. Examples are Stora Kraft's sale of its distribution activities and Korsnäs's sale of its production facilities to Stockholm Energi.

V.C.d. The Finnish patterns
As for Sweden, deregulation of the Finnish electricity market was anticipated by the market actors. As early as 1993, large parts of the Finnish industry started a structural transformation to meet the demands of liberalisation. In the course of 1992 and 1993, PVO, the large industry-owned co-ordinator, consolidated its position as the central co-ordinating core of the heavy industrial electricity supply through taking over ownership shares in several power conpanies previously owned by individual heavy industrial enterprises. The company has also increased the number of shareholders as shareholders in smaller power

companies have sold shares to PVO. In return they have received shares in electricity supply agreements from PVO (Figure IV.11).

At the beginning of 1994, the production capacity of the group was 2430 MW, i.e. 18.5% of the total production capacity of Finland. PVO has in this process increased its ownership share in TVO to nearly 50% and now figures as TVO's largest owner besides Imatran Voima Oy. On the public producer side, IVO already had an entirely dominant position.

The establishment in 1993 of the co-operative group Kymppivoma Oy by ten major distribution companies marked an important step towards matching the increasing producer concentration. The Kymppivoima alliance came into being in a united attempt to secure improved price terms from the country's leading power producer, IVO, and represented the first serious challenge to Imatran Voima Oy's dominating presence in the market. The regional alliance accounts for combined energy purchases of about 10,000 GWh of electricity annually, equivalent to about one-third of electricity consumption in the residential/commercial sector in Finland. However, there has also been a strategic move in the opposite direction. IVO purchased a 28.7% share in Länsivoima Oy, the largest shareholder in Kymppivoima and the second largest distributor in Finland. This gives IVO access to Kymppivoima, through Länsivoima, and through Jyllinkosken, which Länsivoima acquired in hard competition with the Swedish State company Vattenfall.

IVO's active engagement in Finnish distribution is quite new and

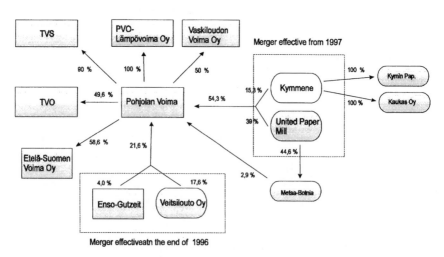

Fig. IV.11. The consolidation of PVO. *Source:* Annual Reports (1994).

was clearly triggered by Vattenfall's aggressive buyout strategy in Finland. IVO started the integration by merging its current distribution activities with another distributor and customer of IVO, Paloheimo Oy (Fig. IV.12). Paloheimo is ranked among the 20 largest distributors in Finland, and the company owned jointly by IVO and Paloheimo, Uudenmaan Energia Oy, will have a distribution of at least 1500 GWh. This is enough to rank the company among the five largest distributors in the country.

V.C.e. Intra-Nordic developments

Strategic acquisitions and alliances in Nordic electricity industries have not only been limited to the domestic market, but have also moved across national boundaries. A power struggle on the Nordic electricity market, with the involvement of major European players intensified in the first months of 1996. The tug of war surrounding the second largest Swedish producer and distributor, Sydkraft, was intensified when PreussenElektra took a 12.4% share in Sweden's Graningeverken one day after France's EDF bought 25% of the company (Fig. IV.13). PreussenElektra, owned by the industrial giant VEBA AG, already holds

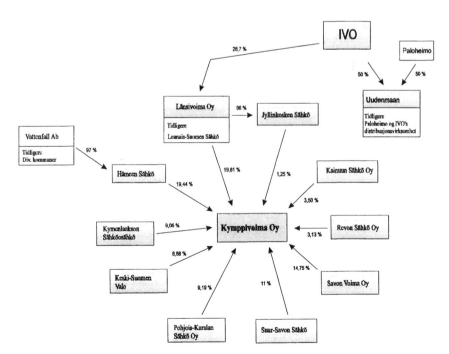

Fig. IV12. The Kymppivoima alliance. *Source:* Eivind Gundersen (1996).

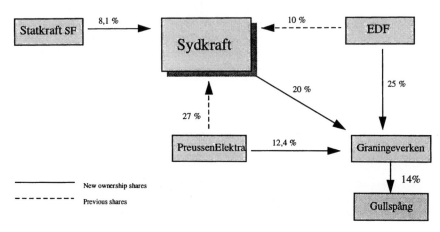

Fig. IV13. The Sydkraft struggle. *Source:* Atle Midttun *et al.* (1996).

17.5% of the capital and 27% of the votes in Sydkraft. With EDF already holding 10% of the capital in Sydkraft, the battle lines have begun to form between European utility companies as they attempt to gain control over major players in the deregulated Nordic markets. The web of cross-holdings between the Nordic region's major power producers is further entangled by the fact that Sydkraft holds 20% of Graninge. Thus, if either PreussenElektra or EDF were to gain control over Sydkraft, Graninge would fall firmly into the German or French sphere. Malmö and two other Swedish cities sold off their combined 5% capital stake in Sydkraft to Norwegian State-owned power company Statkraft, which now controls 8.1% of the voting shares in the company. Some 24% of the capital and no less than 38% of the votes in Sydkraft still remain in municipal hands, and a well-placed bid from either PreussenElektra or EDF could clinch control over a large part of the Swedish market.

In addition to the power struggle between the two large European power companies, there has been another tug of war between Finnish State-owned company, Imatran Voima Oy, and the large Swedish State-owned company, Vattenfall. Imatran Voima has crossed the 50% threshold in the fourth largest Swedish power producer, Gullspång. Shortly after the announcement, Vattenfall purchased a share in Gullspång corresponding to 10.7% of the capital and 6.3% of the votes. The reason why Vattenfall has purchased the shares, considering IVO's 50.1%, is that under Stockholm Bourse rules, Vattenfall's corner holding prevents IVO from taking over Gullspång.

The tension between IVO and Vattenfall is maintained in Finland, where Vattenfall wants to gain a foothold in Finnish distribution.

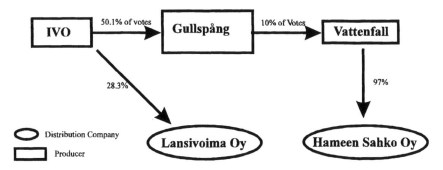

Fig. IV.14. The Gullspång case. *Source:* Eivind Gundersen (1996).

Vattenfall's largest investment is its takeover of Finland's third largest distributor, with a market share of 6%. IVO, on the other hand, has secured its position in Finland's second largest distributors both directly and indirectly through Kemijoki Oy. The jockeying for position in distribution has been intensified with the competing bids the two companies have made for Revon Sähkö, a distribution company with shares in the nuclear producer TVO.

Vattenfall has extended its interest from the domestic and Finnish market to include the Norwegian market. The latest addition to the web of cross-holdings in intra-Nordic power companies is its acquisition of 10.1% of the shares in the Norwegian power producer Hafslund Energi AS. Hafslund Energi was, in 1996, separated from Hafslund Nycomed AS, and is now a separate electricity company listed on the Oslo Stock Exchange. The Norwegian State-owned oil company, Statoil, has in the same time period purchased 12.3% of the company (Fig. IV.15).

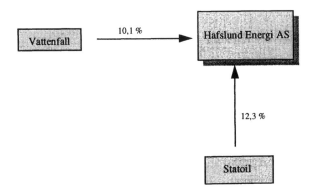

Fig. IV.15. The Hafslund case. *Source:* Atle Midttun *et al.* (1996).

V.C.f. European and inter-continental developments
Following the liberalisation of the Nordic electricity markets, a commercially based trade is developing between the Nordic peninsula and the non-liberal regimes in northern continental Europe. From Norway, contracts have been signed for two overseas cables to Germany and one to The Netherlands, each with an approximate capacity of 600 MW. Additional cables have been laid from Norway to Denmark, adding up to a capacity of 1040 MW. From Sweden, a cable of 600 MW has been laid to Germany; two connections of 670 and 1300 MW, respectively, have been laid to Denmark. A 600 MW cable is also projected to Poland, which is a joint venture in which Svenska Kraftnät will hold 50% of shares, Vattenfall 49% and PSE (Poland's national power company) 1% (Fig. IV.16).

Common to these trade relations is that they take place under mixed institutional regimes, with varieties of liberal regimes on the Nordic side, and varieties of public-service oriented, cartelised, or regimes in negotiated transition on the other. The basic rationale behind this trade lies in the attractive complementarity between Nordic hydro-power and continental heat-based power. In a combined system, hydro-power, with practically no start-up costs, can be used for top-load production, whereas heat power can be used for base-load and dry-year security.

The driving motivation on the liberal Nordic side lies in commercial advantages at the company level. On the continental side, this varies according to regimes, and spans from company-based commercial considerations in the German PreussenElektra case, to also include broader societal interests in the case of public-service oriented companies.

In spite of a common commercial attitude to overseas electricity trade, the three liberal Nordic systems differ significantly in their contractual design. The Swedish model is characterised by strong company control, partly coupled to cross-ownership between the contracting parties. Sydkraft's contract with PreussenElektra regulating the Baltic cable is thus supplemented by the acquisition of share in Sydkraft from PreussenElektra, and a joint investment by the two companies in a coal-based power station in Lübeck. Vattenfall's agreement with Sjællandske Kraftverker, which includes a right to transit 200 MW through the cable between Denmark and Germany, also contains mutual ownership exchange and control of production capacity in each other's systems.

The Norwegian model apparently supplements the commercial regime with a societal orientation through the intervention of Statnett, the State grid company, as owner of the cables from the Norwegian side. Statnett thus wholly owns the 1040 MW cable connection to Denmark, and also owns 50% of Eurocable and Viking Cable and the Dutch SEP cable. The German Euroström, and PreussenElektra and the

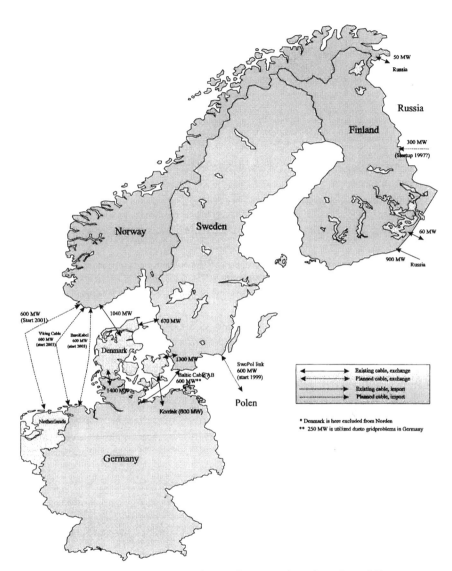

Fig. IV.16. Cable connections between the Nordic peninsula and continental Europe.

Dutch SEP are co-owners with an equal share. Within this context, a trade agreement is then specified between Norwegian production companies and the continental partner.

The Finnish model is characterised by strong integration between the power importer and the governance of the export cable. The 900 MW cable agreement with Russia is a case in point. Imatran Voima and its

Russian counterpart are responsible for investments and run the cable on commercial terms. Imatran Voima will in the near future also take over the Russian exporter, and thus integrate the whole trade as an internal company transaction.

Nordic–continental trade is, as we have seen above, increasingly paralleled by investment strategies into the Nordic market, especially into Sweden and Finland, where concession rules are not as prohibitive as in Norway

V.D. Concluding remarks

Besides the British Isles, the Nordic area has clearly exhibited the most dynamic liberal development in Europe. Nordic liberalisation is of particular interest because it involves collaboration between several countries, and thereby also implies the development of a common international free-trade regime. As such, it may provide an arena for liberal learning as well as a prototype for a further liberal European development.

Compared to the larger European context, the Nordic countries are probably more homogeneous and may therefore more easily arrive at common solutions to the controversial issues of market regulation. Of particular notice in this context is the pragmatic and broadly consensual political approach to the liberal-market issue, which was broadly shared by all countries involved. The pragmatic Nordic consensus is, however, probably reliant upon the electricity market not getting too out of control. Should excessive price increases or British-style superprofits and massive layoffs of staff come to dominate the scene, we would be likely to see political reactions. Reactions would also be likely if the Nordic scene was seen to become too much of an arena for big multi-national power games.

References

Norwegian Parliamentary Documents
The Ministry of Oil and Energy (1985) Energy Law Issues, *NOU No. 9.* The State Printing Office, Oslo.
The Ministry of Oil and Energy (1988–89) The Energy Law, *Government Bill,* Ot prp. No. 73.
The Energy and Industry Committee (1988–89) *Withdrawal of The Energy Law, Parliamentary Treatment of Government Bill, Innst O, No. 93.*
The Energy and Industry Committee (1989–90) The Energy Law, Parliamentary Treatment of Government Bill, Innst O, No. 67.

Other Norwegian sources
Annual reports from the 20 largest Norwegian producers and distributors for 1994.

Energy Law Issues (1985) No. 9, The State Printing Office, Oslo.

Bjørndalen, Jørgen, Hope, Tandberg and Tennbakk (1989) *Market-Based Electricity Exchange in Norway*, Working Paper, Centre for Applied Research, The Norwegian School of Economics and Business Administration, Bergen.

Hope, Einar and Strøm, Steinar (1992) *Energy Markets and Environmental Issues: a European Perspective*, Proceedings of a German–Norwegian Energy Conference, Bergen, 5–7 June, 1991, Scandinavian University Press, Oslo.

Midttun, Atle (1996) Electricity liberalisation policies in Norway and Sweden, *Energy Policy*, Vol. 24, No. 1.

Midttun, Atle (1995) Negotiated adaptation from plan to market, *Sosiologi i dag*, Vol. 4.

Midttun, Atle (1994) *Norwegian Electricity Reform: A Case of Competitive Regulation*, 16th Canadian National Energy Forum, World Energy Council, Montreal.

Midttun, Atle and Køber (1995) *Markedsstruktur og Prisstruktur–Rapport nr. 4 til NVE*, Forskningssenteret, Handelshøyskolen BI, Sandvika.

Midttun, Atle, *et al.* (1996) *Strategi og Struktur i Nordisk Elektrisitet Industri*; 1996. Research report, Norwegian School of Management.

NVE (1993, 1994, 1995): *Electricity Market Surveys*, NVE Publications, Oslo.

NVE (1996) *Price Development for the Norwegian Market*, NVE Publications, Oslo.

Ministry of Industry (1995) *Fact Sheet 1995*, Government Publications, Oslo.

Ministry of Trade and Industry: Fact Sheet 1996.

Statnett Marked (1994) Special Calculations provided by Hans Randen. Trade Director, Statnett Marked.

Statnett Marked (1996) *Country Report No. 1–20*, Statnett Publications, Oslo.

Statnett Marked (1996) *Spot Prices 1985–96*, Statnett Publications, Oslo.

Statnett Marked (1996) *Key Numbers from Country Reports 1990–95*, Statnett Publications, Oslo.

Statnett Marked (1996) *Net Exchange per Week from 1985*, Statnett Publications, Oslo.

Statnett Marked, *Weekly Production and Consumption Reports*, 1990–1996.

Thue, Lars (1995) Electricity rules: the formation and development of the Nordic electricity regimes. In Kaijser, Arne and Hedin, Marika (eds) *Nordic Energy Systems: Historical Perspectives and Current Issues*, Science History Publications, Massachusetts.

Swedish Parliamentary documents

Ministry of Industry (1990–91) *An Industrial Policy for Growth*, Government White Paper, Regjeringsproposition 87.

Ministry of Industry (1991–92a) *Electricity Market Issues*, Goverment White Paper, Regjeringsproposition 124.

Ministry of Industry (1991–92b) *An Electricity Market with Competition, Government White Paper*, Regjeringens proposisjon 133.

Other Swedish sources

Annual reports for the 20 largest Swedish producers and distributors in 1994 .

Bergman, L., Hartman, T., Hjalmarsson, L. and Lundgren, S. (1994) *Den nye elmarknaden*, SNS Forlag.

Fritz, Peter and Springeldt, Per Erik (1995) *Buying Electricity: A Handbook for Exploiting the Open Electricity Market*, The Swedish Confederation of Industry, Stockholm.

Midttun, A. and Summerton, J. (1985) *Loyalty or Competition? A Comparative Analysis of Norwegian and Swedish Electricity Distributors' Adaptation to Market Reform.*

Ministry of Industry (1996) *The New Swedish Market Reform: The Electricity Reform in Shorthand*, The Government Publication Office, Stockholm.

Ministry of Trade and Industry (1994) *Fact Sheet 1994 and NUTEK (Swedish National Board of Industrial and Technical Development)* Energy Statistics.

NUTEK (1991) *The Electricity Market in Change: From Monopoly to Competition*, Report No. 6, Gotab, Stockholm.

NUTEK (Swedish National Board of Industrial and Technical Development) (1995) *Energy in Sweden 1995*, annual fact sheet.

Summerton, Jane (1994) *Changing large technical systems*, Boulder, Westview Press.

Finish Parliamentary documents

Ministry of Trade and Industry (1994) *The Electricity Market Act*, Government Bill.

Ministry of Trade and Industry (1979) *The Electricity Market Law*, Government Bill No. 319.

Ministry of Trade and Industry (1989) *Change of the Electricity Law*, Government Bill No. 59.

Other Finnish sources

Annual reports for the 20 largest Finnish producers and distributors for 1994.

Association of Finnish Electric Utilities (1994) *Electricity Statistics for Finland, 1994, Helsinki*.

Gundersen, Eivind (1996) *Strategic Interaction in the Finnish Electricity Market*, Masters Thesis The Norwegian School of Management, Oslo.

Ministry of Trade and Industry (1994) *Finnish Energy Review 4/94*.

Reuters Business Briefing (1993) *Electricity companies accused of slowing down deregulation* (November 15).

SOM/EL-EX Electricity Exchange Ltd (1996) *Finnish National Electricity Exchange Will Start*, Press release, February 16.

Other Nordic sources

NORDEL (1994) *Årsberetning*, NORDEL's Sekreteriat, Skjellandske Kraftverker.

Nordic Council of Ministers (1994) *Elhandel med lande udenfor Norden*.

Reuters News Briefing: July 5, 1993; November 13, 1995; April 28, 1996; March 14, 1996; April 18, 1996; April 19, 1996.

Tema Nord (1994) 626 Nordiska ministerådet.

OECD (1995) *Main Economic Indicators* (1960–1996), Paris, OECD.

IEA (1994) *Energy Statistics of OECD Countries* (1960–1994), OECD/IEA, Paris, OECD.

Part Three

Systems in Negotiated Transition

Chapter V
Energy Efficiency and the Political Economy of the Danish Electricity System

FREDE HVELPLUND

I. Introduction

I.A. Background

Environmentally defined 'public service' requirements have, in recent decades, increasingly been levied on the Danish electricity power system. In 1990, for instance, the Governmental energy plan 'Energy 2000' (Ministry of Energy, 1990) introduced the aim of reducing Danish CO_2 emissions by a minimum of 20% before the year 2005 and by 50% before 2030 as compared to the situation in 1988. In 1996 a new Governmental energy plan, 'Energy 21' (Ministry of Environment and Energy, 1996) maintains these aims. Technologically these CO_2 reduction demands require a rearrangement of the electricity system from large power plants towards an increased exploitation of energy-conservation technologies, renewable energy resources and co-generation of power and heat. We call these changes 'radical technological changes', because they not only include changes in techniques, but also in organisations, knowledge backgrounds, etc. Often they cannot be developed and implemented by fossil fuel companies, as they will often lose market share, if the official environmental goals are to be achieved. The new technologies are decentralised and vary from place to place, which requires development-oriented regulation regimes.

The changing European electricity market regimes are, at the same time, putting a new kind of pressure on power companies and public regulation regimes. Consequently, in May 1996 a law was passed by

Parliament, accepting a certain right for some consumer groups to buy electricity at the market.[1]

The Danish regulation praxis could be called 'challenged negotiated regulation', where the administration tries to find solutions in close negotiations with the old fossil-fuel based supply companies, and where the results of this process are persistently challenged by grass-roots groups, independent researchers, renewable-energy interest groups, and vigorous public debate.

In the late 1970s grass-roots anti-nuclear movements were able to hinder nuclear energy and establish a political base for renewable energy. In the 1980s this was supplemented by the rise of a new industry, based on renewable energy and co-generation technologies. Danish regulation can therefore be described as a dialectical process between an administrative and political tradition of negotiated regulation, and a strong political force consisting of the new energy industries, grass-roots movements, a critical press and public opinion. Decentralised co-generation (based on biomass and natural gas, and windpower) and some electricity conservation measures have been introduced in spite of the opposition of the large power plant companies, and as a result of the active political participation of the general public.

I.B. Purposes of this analysis

The growing conflict in the above-mentioned dialectical process is:

1. that we have reached a stage where the new green technologies cannot be developed sufficiently merely by continuing existing regulation procedures and by regulating existing organisations. New organisations and regulation procedures have to be established, which by their internal interest and dynamics develop and improve the new technologies. At the present stage of technological development the proponents of the new green technologies have become influential challengers of the ongoing development, but still need to become more powerful as creators of new technologies and developments.

2. that the 'negotiation' dimension does not work under the new European electricity market regimes, where the negotiation partners will be many, and often undefined, as they are not any longer only Danish supply companies, but also Preussen Electra, Stattkraft, Vattenfall, EDF, etc.; and

---

[1] In this law, the 'liberalisation' is limited by the first priority of producing electricity from renewable energy and co-generation plants. One could also say that liberalisation has been extended by removing existing barriers to entry against new technologies.

3. that a tradition of 'challenged negotiated regulation' leads to over-capacity in a process where the proponents for the old fossil-fuel technologies are getting what they want, and later the proponents of the new technologies will get what they want.

As a result of these changed conditions of regulation, the Danish Government is preparing a new regulation regime based on detailed legislation, which on the one hand is opening the market for limited competition, and on the other hand is protecting environmental public-service obligations and its technologies (co-generation, renewable energy and electricity conservation).[2]

It is the purpose of our analysis to discuss the existing Danish regulation model and its ongoing changes. We also wish to present recommendations regarding necessary alterations in existing and planned changes in the Danish regulation praxis, taking the above-mentioned problems into consideration. We think that a change from 'challenged negotiated regulation' to what we call 'regulation for technological change' is a short characterisation of the necessary reforms. Technological change implies a focus upon improving creative abilities and implementative power of an array of new actors on the energy scene. This implies the creation of new organisations and a change of the old organisations at the energy service supply level, and also at the public regulation level. These new organisational constructions cannot be described in crude terms of liberalisation, oligopoly, public ownership, etc., as they have to be designed as a response to concrete organisational starting points. Therefore, our regulation proposals are responsive to specific organisational conditions in one specific country. Thus our analyses have to be detailed and specific. At the same time, we have to manage new European market regimes, where negotiation becomes increasingly difficult and has to be replaced by forms of legislation. That is why our proposed regulation regime is often also legislative.

The questions which will lead our investigations are the following:

Question 1. Which type of dynamics is built into the organisation of the Danish direct electricity supply system?

Question 2. Which dynamics are built into the Danish democratic system regarding the ability to regulate the Danish electricity system?

Question 3. Which dynamics are built into the combination of electricity system and democratic system regarding their ability to establish regulation procedures between the new

---

[2] See 'Lov om ændring af lov om elforsyning', which was passed by Parliament on May 31, 1996.

European market regimes and the environmental 'public service' goals?

II. The Danish Electricity System

II.A. Consumption, environmental effects and supply technology

Danish electricity consumption in 1994 was 31.2 TWh, 18.2 TWh being consumed in the Jutland–Funen region and, 13.0 TWh in the Zealand region. Between 1981 and 1994, electricity consumption increased by about 40%. According to use, consumption can be classified as follows: households, 31%; industry, 29%; commercial and service, 28%; agriculture and market gardening, 8%; street lighting, electrified transport, etc., 4%. If the category 'large enterprise' is defined as those consumers using more than 1.5 GWh *per annum*, consumption by all large enterprises was 0.8 TWh in 1990, or less than 3% of total consumption.

In the 1980s, the resource base for the electricity system was coal, representing more than 95% of fuel consumption. In the 1990s this is changing in the direction of biomass and natural-gas based co-generation units. In 1996, windpower and combined heat and power plants (CHPs) deliver 27% of the electricity consumption in the Jutland–Funen area, *versus* 1% in 1987 and 4% in 1991.

Annual fuel consumption for electricity production at the present time is between 9 and 11 million tons of coal. The variation is caused by a variation in the annual amount of imported electricity, especially from Norway. In terms of environmental effect, SO_2 emissions (corrected for import) has decreased from about 200,000 tons *per annum* in 1981 to about 89,000 tons in 1994. NO_x emission has decreased from 100,000 tons *per annum* in 1981 to 78,000 tons in 1994. In the same period (1981–1994) CO_2 emissions from electricity production have increased from 23 million tons *per annum* to 28 million tons *per annum*. CO_2 emissions from the electricity system represent *c*. 50% of total CO_2 emissions, in Denmark, including that from transport. Therefore, the development of the Danish electricity system is of great interest in achieving the Danish energy policy goal of reducing CO_2 emissions by a minimum of 20% in the period from 1988 to 2005.

II.B. Organisation of the direct electricity supply system

The direct[3] electricity supply system in Denmark is defined as the power, transmission and distribution system. The organisation of the direct electricity supply system[4] is divided into two main regions; ELKRAFT, which is a co-operative covering the area of the island of Zealand and neighbouring islands; and ELSAM, which is a co-operative covering the area of the Jutland peninsula, the island of Funen, and neighbouring islands.

The electricity system is in principle hierarchic ('bottom up'), which has a formal direct consumer ownership (co-operatives/limited companies) in the case of 53% of the distribution companies, and indirect consumer ownership owned by municipalities for the remaining 47%. The latter form is common in Danish cities. The 'bottom' (the consumers) delegate power to a hierarchical power system, power which is controlled by open information and election procedures, where the consumers elect the board of directors at the different levels of the hierarchy. In the ELKRAFT area only a minority of the distribution companies are co-operatives with direct consumer ownership; municipal ownership is the rule. In the ELSAM area, 30 out of 81 distribution companies are owned by the municipalities, and the rest (51 companies) have direct consumer ownership *via* a co-operative organisation.[5] The 51 distribution companies organised in direct consumer ownership account for 67% of electricity sales in the Jutland–Funen area. The Jutland–Funen power system has six power companies, each owned by the distribution companies in their region. The largest, Midtkraft, produces 22%, and the smallest, Skærbækværket, 9% of Jutland–Funen electricity sales. The 81 distribution companies in Jutland–Funen each have their area monopoly, the largest selling 700 GWh/year, and the smallest, an island, 1 GWh/year. Regarding the distribution companies owned by the municipalities, the representatives are elected by the members of the Town Council. Regarding the consumer-owned co-operative distribution companies, consumers elect their representatives in the distribu-

---

[3] The *direct* electricity receiver system is defined as the apparatus transforming electricity to electricity services, plus the organisations organising this process. The *indirect* electricity receiver system is defined as the firms and infrastructure supplying the apparatus and organisational knowledge into the *direct* electricity receiver system.

[4] As mentioned above, a new electricity production system has been evolving since the 1980s. 27% of the electricity production in Jutland–Funen is now coming from decentralised co-generation plants and windpower. These systems are mostly organised as consumer-owned co-operatives.

[5] This includes ten distribution companies organised as private foundations with consumer election of representatives and the board of directors.

tion companies. These representatives elect their board of directors. This board of directors elect their representatives for their power-plant company. The director and the deputy director of each power plant are automatically members of the power-plant association for Jutland–Funen (ELSAM). ELSAM is the co-ordination unit and the organisation which elaborates the political strategy for the electricity system in Jutland–Funen (see Fig. V.1).

The Danish electricity system is usually described as a non-profit system, because the Electricity Supply Statute does not permit the electricity companies to build up a surplus, or to use any surplus for non-energy purposes. An accumulated surplus in any year must be returned to the consumers *via* lower prices in the following year. However, the system is profit-driven, in as much as it is owned and administered by the consumers, who receive the rewards from its cost-effectiveness. From this point of view, we find it better to describe the Danish electricity system as a consumer-profit system.

To a great extent, the electricity system is self-financing, as the consumers pay in advance *via* the electricity prices in a five-year period before a new power plant is ready for production. Due to the fact that the electricity system is an old, well-established system, it is so well consolidated that debt is limited to the value of its coal stocks, and the power plants and transmission and distribution lines are without debt.

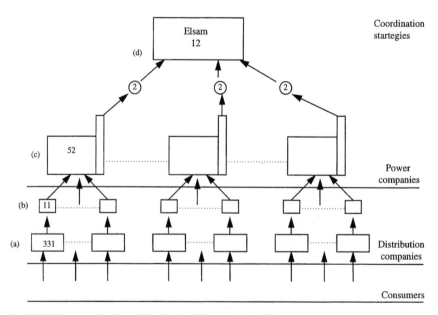

Fig. V.1. The organisation of the Jutland–Funen electrical power system.

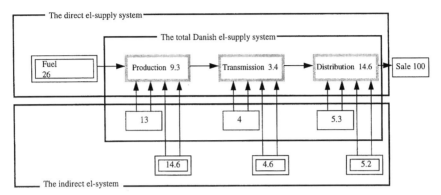

Fig. V.2. Allocation of the added value of the whole electricity supply system in Denmark. *Source:* Calculated on the basis of statistics contained in SØ89-112, April 10, 1989, ELSAM; Statistic 1991, Dansk Elforsyning, DEF; and Elforsyningens tiårsoversigt 1980–1989, DEF.

II.C. *Value added and horizontal and vertical integration in the Danish electricity service supply system*

The allocation of the added value of the whole electricity supply system is structurally described in Fig. V.2,[6] which shows how delivery of electricity for 1 Dkr at consumer level is distributed at different levels of the value-added chain. In 1989, this represented a supply of 2.22 kWh to the consumer. Figures framed by a double line in Fig. V.2 represent the import, and amount to a total of 50.4%/øre. Figures framed by a grey hatching in Fig. V.2 represent the added value related to the employees in the electricity system, external assistance to operational expenditures and from the public infrastructure related to these activities. They amount to a total of 27.3%/øre. This figure is defined as the added value in the Danish part of the direct electricity supply system.[7]

---

[6] The principles used in the calculation, because of the fact that these divisions are not used in the 10-Year Summary ('Tiårsoversigt'), are that the 1989 figures are based on Table 25 of the Summary, and the allocation of costs between production and transmission plant and constructions, are based on figures contained in SØ89-112.

[7] It should be explained that the increased value of the Danish electricity system represents the wages of its 11,600 employees. The employees' ability to produce the increased value is a result of the education and training they have received, prior to their employment in the system. This means that these costs should be deducted from the increased value of the electricity system because the educational system is a subcontractor of knowledge, attitudes, etc., to the electricity system. This thought may be extended, because the civil society's education and care of the labour force of the electricity system is that part of the indirect electricity system. Thus, the direct system is in reality only a *purpose* (electricity generation and distribution), and a *structure* (organisation of electricity supply), which organises and directs the indirect electricity generation's factors of production. In this chapter, in order to perform the analysis, it has been decided to define the wages of the employees as a part of the direct electricity system's increased value.

Table V.1. Allocation of an electricity supply of 100 øre to the consumer, on plant and operating costs, and on import to and production in Denmark

Fuel imports	26.0 øre
Plant import (indirect electricity system)	24.4 øre
Danish plant (indirect electricity system)	22.3 øre
Danish electricity operation costs (direct electricity system)	27.3 øre
Total	100.0 øre

Table V.2. Allocation of a refrigeration service supply costing 100 øre at consumer level

Fuel imports	5.9 øre
Plant import (indirect electricity system)	5.6 øre
Danish plant production (indirect electricity system)	5.1 øre
Danish electricity production and distributions costs (direct electricity system)	6.2 øre
Refrigerator costs	77.0 øre
Total refrigeration costs	100.0 øre

Figures framed by a single line in Fig. V.2 represent the Danish added value of the expenditures for plant, etc., and therefore the Danish part of the added value in the indirect electricity system. Together this amounts to 22.3%/øre. To summarize, it can be said that the added value resulting from the production of electricity for 100 øre (1 Dkr) is allocated as shown in Table V.1.

When power-generating statistics are examined, and likewise the number of employees in the electricity sector (production, transmission and distribution), it can be seen that 27.3% of the price of electricity represents wages paid to the employees in the electricity sector, and to subcontractors of services to the electricity sector's operations in Denmark.[8] It is also interesting to look at this value-added chain from an energy service point of view, where the goods are energy services and not kWh. In Table V.2 the value-added chain for refrigeration[9] at 7°C is illustrated. This example points out that it is worthwhile to remember that the electricity sector is only a very small fraction of the energy service supply sector. The Danish sector is especially small, because it has no integration with fuel production and plant production.

---

[8] The allocation of added value is made under the assumption of the present conditions of financing in the direct part of the Danish electricity system, i.e. the interest rate is approximately 1% *per annum* when the investment is in new capital. The electricity price which is the basis of the calculation contains an interest-rate expenditure of approximately 1% *per annum*.

[9] Refrigeration capacity of 200 l with a yearly consumption of 200 kWh. Refrigerator price 3000 Dkr, and depreciation time 10 years. The interest rate used is 5% *per annum*.

III. The Dynamics of the Danish Direct Electricity Supply System

The most interesting question for our purposes is the dynamics of the electricity system with regard to capability of change, inherent growth dynamics, cost regulation and ability to influence Parliamentary processes.

As shown above, the Danish direct[10] electricity system is a relatively small system providing only about 27% of the added value, which is represented in the price of the electricity.[11] Co-ordination within this system is not regulated by conditions of a traditional competitive market.[12] In this part of the direct electricity system, the Danish system is vertically integrated and influenced by agreements made on the basis of a network of reciprocal ownership relationships. Historically, it is a very democratic organisation. Democratic election to the bodies responsible for transformer stations and distribution lines ensure consumer control. At the beginning of this development, when the plants were small, this control functioned well. The situation has now changed. This is the result of the development of the technical organisation towards larger and larger power plants and a larger electricity-transmission system. This development has resulted in decreased consumer control and increased control at director level in the power companies.

The organisation's combination of openness with regard to pricing, the system of profit repayment to the consumers and consumer ownership and management, appear to ensure a high degree of cost effectiveness in the electricity system. The hierarchic structure of the system, with a series of indirect elections, in combination with the dominance of the power interests at the top of the system, results in a technological conservatism which is bound to a certain centralist and supply-orientated paradigm. The fact that there is no lobbying from coalminers and stock-holders, and that the electricity system is very consolidated, makes it possible (although still difficult) to establish public regulation regimes, when necessary, against the interests of the power companies.

---

[10] Consisting of power generation, transmission and distribution.

[11] The German system, for example RWE Energie, which owns lignite mines and the subcontractors who deliver the power plant, organise 60–80% of the added value in the final price for the electricity.

[12] There is no competition in the sense that a dissatisfied customer of an ineffective and expensive distribution company can choose another supplier. There is competition in the sense that the dissatisfied customer can compare the price of electricity from his or her distribution company with the corresponding price in another company. Thereafter, the dissatisfied customers, as owners of the electricity system, can put pressure on their consumer representative, who has the ability to change the management of the company.

III.A. The Danish direct electricity system and the capability for technology change

The consumer representatives at the ELSAM/ELKRAFT level are not in harmony with the citizens' desire for more environmentally friendly electricity production. In terms of the ability to incorporate radically new technologies, the organisation is far from being competitive. Minority interests are excluded in the series of indirect elections. Because new ideas, by definition, almost always are minority ideas, they do not find a way through to ELSAM's board of directors. Furthermore, ELSAM's board comprises the chairman, vice-chairman and managing director for the seven power plants, and thus represents a certain technology, i.e. centralised power plants. In principle, new ideas may be pushed through the hierarchy of the electricity system, but other channels are quicker and easier to use. Thus, the democratic channels of the electricity system are bypassed by a combination of popular movements and political channels through district, regional and national public authorities and institutions.

In a situation where there is a need for radical change of technological direction, the form of the organisation of the power plants creates conflicts in relation to the rest of society. Their organisation is geared towards the efficiency norms of the 1950s and 1960s concepts of 'expansion and centralisation', but times have changed. The norms of today are 'cost reductions and decentralisation'. The power companies have difficulties in developing new ideas. Therefore, there is a conflict in relation to parliamentary democracy, which still has this capability to a certain degree.

It is important to note the following features:

1. The elected board of directors of ELSAM has a preference for large, centralised power plants (at present, coal plants). This is the case because the board consists of the elected director and deputy director in each of the six large coal-based power plants. There is no independent administration linked to the elected board of directors.
2. Furthermore, the administrative directors of these companies have the right to meet (but not to vote) at the 4–6 committee meetings every year. There is no independent administration linked to the elected board of directors.
3. A set of indirect election procedures squeezes out any minority group, and new technology is always a minority idea at the initial stages of its development.
4. There is a high degree of openness with regard to accounting and tariff information. Every year information regarding these matters

is published. That way, any consumer can see what other consumer groups and consumers in other areas pay for electricity. This public control makes the Danish power companies rather cost-efficient. Denmark has the lowest electricity prices in the EU, after Sweden and Finland.

5. The Danish electricity system is consumer-owned, and it is stated by law that any payment for electricity has to be recycled to the electricity consumers. This system has (among others) the result that the Danish electricity system in reality has no debt. It can, for instance, lose market shares without risking any serious economic problems.

Points 1–3 above result in an organisation which is very conservative concerning technology, and the conserved technology is a centralised coal-based system. Therefore, it is no wonder that this type of organisation has worked against the introduction of decentralised co-generation plants since 1975. Points 4 and 5 above result in an organisation which is rather cost-efficient and governable, due to relative openness and the financial ability to survive changes. All these characteristics are important when analysing this organisation as a part of a process of change. One could say that the Danish electricity organisation cannot change itself, but it can survive in a process of change.

There is no taxation on power-company profit, because, by Statute, power companies must pay back profit to the consumers. No capital tax is levied on company assets, which are calculated to be approximately 40 billion Dkr. No concession royalties are levied.

III.B. The direct electricity supply system's motivation for increasing the electricity consumption

III.B.a. Expansion on account of the dynamics of co-ordination
In the power companies, there are large groups of employees whose employment is dependent on the construction of new plants and on the future of the power companies. Every one of these companies is represented in ELSAM's board of directors, and is protected from abandonment by a series of agreements. The finely balanced system of co-operation, with complicated volume and price agreements on mutual power delivery, functions without market competition. It is a mutual support system, where the individual power company cannot 'go bankrupt'. Within ELSAM, all of the six power companies have a desire for plant extension. The costs are mutual, but they cannot all be extended at the same time. In 1991 Vestkraft Power Generation Capacity (PGC) in Esbjerg had just been extended, just after Funen PGC in Odense, and

the following on the list were Skærbæk PGC and Nordkraft PGC. Expansion in the form of more power plants is a necessity in the ELSAM operation. The proverb 'the horse bites when the crib is empty' is very valid in an organisation of this type.

ELSAM wishes to implement a long-term plan for a high-tension grid for the entire Jutland–Funen area. The aim of this network is to provide a second, relief, high-tension supply to all parts of the area. The largest capacity is now 400 kV, which ELSAM describes as the power motorway. The area already has several supply routes, but only at 150 kV. When the more powerful grid is completed, ELSAM will implement a more effective daily reorganisation, in which they will own the 400 kV network, whilst the 150 kV network will be transferred to the ownership of the individual power companies. At the present time, only four or five of a total of 15 sections need to be completed in order for the high-tension grid to be accomplished. The section between Aalborg and Aarhus is one such remaining section.

ELKRAFT has similar plans. On Zealand, the expansion is concentrated on one large power company, Sjælland PGC, which supplies 85% of the generated power within the area. The recent expansions have resulted in an agreement for 350 MW to be sold to the former East German electricity company VEAG in the period 1996–2006.

At the same time, ELKRAFT is applying for permission to build a new power plant, Avedøre II, with an installed effect of 460 MW, starting in the year 1999/2000, using the argument of insufficient capacity. This latter station is being advertised under the label 'multi-fuel concept', because a minor portion of the fuel supply can be provided by straw and wood chips. Half of this predominantly coal-fired power plant is planned to be sold to the Swedish power company Vattenfall, in return for the purchase of a 200 MW power capacity in a Swedish hydro-electric power station.

III.B.b. Expansion on account of the dynamics in the allocation of internal costs

The electricity system's regulation on capital transference is favourable towards electricity distribution companies with an expanding consumption. The old solidarity principle from the initial electrification period, when those who had been provided for were prepared to pay for continued electrification, is still valid in this regulation. The distribution companies which make an effort to save electricity do not pay less in capital transference levies per kWh to a new power plant than a distribution company which does nothing towards saving electricity.

III.B.c. Expansion on account of the dynamics in the allocation of external costs

The consumer who economises on electricity consumption pays the same in capital transference levies per kWh to a new power plant as a consumer who does nothing towards electricity saving, and thus increases the need for a new power plant.

III.B.d. Expansion on account of democratic influence

Democratic influence at both power-plant level and ELSAM level is dependent on the amount of electricity used. This means that decreased electricity consumption in a distribution company results in decreased influence at the power-plant and ELSAM levels.

III.C. The Danish direct electricity system's cost efficiency

Consumer ownership, openness of information and non-profit legislation ensures a cost efficiency which appears to be comparable (and may be even better) than cost regulation in a system where the consumer can select between different power producers. Cost regulation operates by means of openness[13] about and publication of consumer prices and costs, in combination with the consumer representatives' interest in holding costs low. In this area the consumer representatives safeguard a common interest for the consumers, even though they are elected in a process where the voting participation is only approximately 2%.

If a shareholder-owned system is compared with a consumer-owned system, the following can be said: in a shareholder–profit–dividend system, the interest is in maximising profit by means of a combination of high electricity prices and low costs. The idea in a shareholder-owned system is that the enterprises must compete with one another for shareholder capital. By means of this competition, capital moves towards the activities which provide the largest surplus. The cost of such a system is the payment of dividends to shareholders for their investment; the secondary effect is that the shareholders have an interest in a large profit. This may be achieved by a combination of higher prices and lower costs. Therefore, a shareholder-owned system may have an interest in bribing and lobbying the public sector, in order to achieve a

---

[13] It is interesting, in this connection, that the Danish electricity system is based on co-ordination, plus a high degree of openness of information. The English system, following privatisation, is based on the market, plus a very modest degree of openness of information. It is a matter of conjecture as to which of these two systems is closest to the free market's utopia.

monopoly and hinder a public price control. It is this mechanism which operates in Germany, where the electricity companies pay concession royalties to regional and district local authorities, and where these authorities continually enter alliances with the electricity companies in order to prevent greater competition on the market.[14] Hitherto these alliances have been successful, in as much as they have successfully prevented attempts to establish greater competition on the German electricity market.

In theory, in a consumer-owned and consumer-shared profit system, the interests are low costs *and* low electricity prices. There will be no interest in higher electricity prices, as opposed to the interest in a shareholder-organised enterprise. Therefore, there will be no interest in bribing the public price regulator.

In a system with area monopoly (such as the Danish system) the motivation for cost efficiency and low electricity prices could be eliminated by monopoly inefficiencies and the growth of an expensive bureaucracy. To avoid this, alternative control mechanisms to market competition between different power suppliers have to be introduced.[15] In the Danish case, this alternative control mechanism has been successfully implemented as a combination of openness about tariffs and sales conditions, and an effective control by means of the consumers' elected representatives. Both forms of organisation will continually, each in their typical way, attempt to influence the economic framework to their own benefit. The matter of which system will function better, in relation to given societal aims, is completely dependent on the detailed relationship between the electricity system and the existing political system.

III.D. International price competitiveness

The Danish electricity system has the third-lowest average electricity prices in the EU. Only Sweden and Finland have slightly lower prices.

---

[14] See, for example, *Energiekonsensusrunde Ost* of March 31, 1996, where the power companies obtained the permission of the Bundeswirtschaft Ministerium to use cross-subsidiation in order to expand the monopoly. The text is available at the Bundes Ministerium für Wirtschaft und Industrie in Bonn.

[15] It must be kept in mind that the 'liberalised' systems (such as those in Sweden and the United Kingdom) do not have many independent power suppliers. Therefore they cannot be called liberalised systems without market conditions close to what is called 'perfect competition' in the economic theory sense. It can be argued that the Danish regulation regime is closer to 'perfect market' conditions, as the regulator is systematically removing the 'barriers to entry' for new technologies. The result of this policy is that production from decentralised co-generation plants and windmills has grown from around 4% in 1990 to 27% in 1996.

The average Danish electricity prices for industry are around 50% of the German level. Small industrial consumers and farmers, who are typical Danish consumers, are paying three times as much in Germany as in Denmark. The price structure is characterised by the lowest price differentiation between large and small consumers in the EU, and probably in the whole world. Small industrial consumers and farmers even pay lower prices than in Sweden. Furthermore, the capacity payment is lower in Denmark than in any other EU country. This price structure is closely linked to the culture of co-operative ownership, where a precondition for fruitful organisational collaboration is that there is no large difference in the prices paid by different consumers.

III.D.a. Competitiveness in the home market

It is possible (but not probable) that new electricity reforms, making it possible for the distribution companies and six large companies (with an annual electricity consumption of more than 100 GWh) to buy from other producers than the Danish power plants, will result in less power production in Denmark. However, the Danish electricity system has no debt, and is independent of the development of the capital market. Even if the power plants from one day to the next should lose 25% of their market, the price would only increase by 3–5 øre/kWh, or 10%. In such a scenario, the power-plant capacity would, on a long term basis, be adapted to the new situation, and the initial 3–5 øre/kWh price increase would fade out.

The Danish power-plant companies can afford to lose markets, as they have no debts. At the same time they are financed by the distribution companies, and if these companies find it cheaper to buy from foreign companies, this means cheaper electricity on a long term basis. At present, an adaptation to the changing regulation regime has been prepared in the new electricity law. The power distribution companies have obtained increased freedom to buy electricity from power plants other than their own.

III.D.b. Competitiveness in the foreign electricity markets

At present it is not possible for the Danish power companies to use the capital collected in the home market as a base for electricity export and investments in other countries. In that sense, the Danish power companies have a handicap in relation to PreussenElectra, Electricité de France, etc. However, there is no hindrance to Danish power companies establishing independent limited companies for investment outside Denmark. This is possible if the money of the electricity consumers is not used.

In conclusion, in relation to price competition, there is no real danger of losing the home market. So far it is not possible to use the money of electricity consumers as a base for electricity export, overseas investment or investment in new products. The 'no debt' situation makes it very difficult for foreign companies to sell electricity on long-term contracts at lower rates than can be offered by the Danish power companies.

III.E. International-market power competition

It should not be forgotten that a liberalised market in economic theory means (among others) that there are (1) many independent buyers, (2) many independent sellers, and (3) a situation where companies have no influence upon the institutional setting of the market place. At present there is a tendency to focus upon the 'independent buyer' dimension without any serious analysis of the process of developing market power and corporate political power. Nevertheless the real fight on the electricity markets is a fight on market power.

Although there is no free market at all in Germany, France or the United States, companies from these countries are at present buying parts of the electricity systems in countries which have 'liberalised' their markets. At present, around 25% of the distribution companies in the United Kingdom are owned by North American companies. PreussenElectra has bought 35% of the former East German company VEAG, and is now strengthening its monopoly power in the former East German area, together with the other electricity companies from the former West Germany.[16] At the same time, PreussenElectra has bought more than 20% of the Swedish company Sydkraft. The state-owned monopoly company Electricité de France has bought 10% of Sydkraft, etc.

The reality (the threat) is not price competition, but that foreign power companies are very interested in buying the Danish electricity system. This is why there is an ongoing discussion regarding the organisation of the electricity system. At the beginning of 1996 a consultancy firm (PA Consult) submitted a report to ELSAM, advising that the power companies should be changed to joint-stock companies. This proposal is at present being discussed, and has met with arguments

---

[16] See *Verhandlungen zur 'Energikonsensrunde Ost'*, which was agreed upon on 31 January, 1996. In this agreement, the Minister of Industry, Günther Rexroth welcomes that the German power companies have agreed upon using 'cross subsidisation' to establish 'barriers to entry' against electricity produced at independent co-generation units and against electricity from other nations.

regarding the threat of being bought by foreign companies, for instance PreussenElektra. At present the Minister for Environment and Energy supports continued consumer ownership, and does not seem to accept any foreign takeover. In 1996, Parliament accepted a resolution supporting continued consumer ownership.

Regarding the market power question, the following conclusions can be drawn.

1. The Danish market is, to a comparatively large extent, protected against foreign takeover when booking at the consumer owned parts of the system.
2. This protection is at present actively supported by the Danish Government.
3. Specific interests at the director level could work for a 'joint-stock company' model, paving the road for a future foreign takeover.
4. On sealand, there is a high proportion of 'joint stock' ownership with the municipalities as stock owners. This makes the sealand electricity systems easier to sell to foreign power companies interests. In October 1996 there are serious rumours indicating that the largest distribution company, Nesa, with 45% of the electricity market on sealand (5.1 Twh/year) can be sold to Vattenfall or PreussenElektra.

III.F. The Danish direct electricity supply system's political dirigibility

Strangely enough, the political dirigibility of large companies has not, in general, been analysed when discussing the future organisation of the power system. We do not find it necessary to analyse organisational suggestions as if they all had the same features regarding their political dirigibility. Especially in situations where technological changes are needed, one should be aware of maintaining/developing organisations which can be regulated by the Parliamentary process.

The above-mentioned deficiencies in the capability of the direct electricity system for technological innovation are not necessarily a catastrophe, in as much as this may be able to be compensated for by the parliamentary system taking over governance. If the combination of public regulation and the direct electricity system has the ability for technology innovation, an electricity system which is technologically conservative is not necessarily a problem. Therefore, it is interesting to see whether the electricity supply system is politically dirigible. This question will be examined in the following. Political dirigibility is determined in the boundary zone between the direct electricity supply system, the indirect electricity supply, and the democratic process.

III.F.a. Conditions which facilitate political dirigibility
The Danish direct electricity supply system's low degree of horizontal and vertical integration. The Danish direct electricity supply system is characterised, as illustrated above, by a relative independence of horizontal and vertical ownership integration, as opposed to Germany and England, where coal workers demonstrate for new coal-fired power stations.

Immaterial profit-related connection to the public coffers. There is no material profit-related connection between the existing centralised electricity supply structure and the public accounts, i.e. a reduction of electricity consumption in the centralised electricity system, by means of the establishment of independent producers, etc., does not, as in the German electricity system and the Danish natural gas system, result[17] in loss of income to the public coffers.[18]

The Danish electricity system is a self-financing non-profit system, which has no debts. There are no capital owners whose share values drop when the electricity supply market becomes smaller. At the same time, this means that the system is without debts, and that it has no real economic problems in connection with an actual fall in consumption. Therefore, the Danish electricity system, in comparison with the German system, cannot be threatened by falling electricity production.

The Danish electricity supply system is subject to demands on openness with regard to prices and costs. This means that, to a relatively large extent, it is possible for participants in the democratic debate to examine the nooks and crannies of the electricity supply system.

III.F.b. Conditions which make political dirigibility difficult
Short-sighted advantages in the construction of a power plant, which binds district and regional politicians. The construction of a power plant, at a cost of 3 billion Dkr, provides employment for the local labour force in the period of construction and afterwards. ELSAM's payment regulations are so formed that it is the power company which has the newest and most efficient power plant which also has the lowest electricity prices. Furthermore, electricity consumers, for example in Jutland–Funen, make a capital transference to finance the next power plant. This means that all consumers pay towards the new station, representing 3

---

[17] There is an electricity levy, but the income from this is not reduced by adopting decentral co-generation.
[18] It is important to realise that while green levies are playing an increasing role for the public finances, there can be a built-in unwillingness in the public administration to work with resource-saving steps which reduce income.

billion Dkr of regional investment, which no regional politician can refuse.

Effect on employment. The immediate effect on employment results in support from the trade unions—especially the metal workers—when the power companies apply for permission to construct a new power plant. It appears to be of importance in the political process that employment in relation to the construction of a power plant is assigned to workplaces which identify themselves as energy-related workplaces. The alternative to a new power plant is electricity saving, decentral co-generation plants, etc. The workers in these alternatives are dispersed, employed in multi-purpose organisations (electrical, plumbing and heating services, etc.), which are generally less aware of their roles in the alternative energy game, and more weakly organised in terms of trade unions.

The electricity supply companies' power position in the decision-making process in the central administration. The power companies have been and still are strongly represented in the decision-making processes in the Ministry of Energy, and in the Energy Directorate. Thus, the interests of coal technology have been taken care of in the 1970s and 1980s. The new technologies did not and still do not have similar political and administrative opportunities and advantages. The so-called 'Committee of Directors' is a good example of this influence, in which the directors of ELSAM and ELKRAFT hold regular meetings with senior staff from the Environmental Agency, the Energy Directorate and the Ministry of Energy. There are no recorded proceedings from these meetings. The Electricity Pricing Committee is a second example, with 50% of the members being from the electricity sector, and no representation from environmental interests or energy-saving technologies. The Electricity Prognosis Committee is a third example, where seven out of 12 members are from the power companies.[19] The Electricity Supply Statute is a fourth example. The Energy Directorate is only an approving authority, with the power to approve or reject. The power companies prepare the agendas, and decide what is to be applied for. The Electricity Strategy Committee is a fifth example. It consists of members from the electricity sector and the central administration and holds a number of meetings, the recorded proceedings of which are not publicly available.

III.G. Conclusion regarding the direct electricity system

The system is relatively cost effective because of (1) the system of consumer ownership and returned profits, and (2) the relative openness on prices and costs.

---

[19] The role of this Committee is described by Lund and Hvelplund (1994).

The system is technologically conservative because ELSAM and ELKRAFT are responsible for the long-sighted development plans. The chairmen and vice-chairmen of the large power companies sit on the boards of directors of ELSAM, and the managing directors of their power stations also have the right to participate in board meetings. Thus, ELSAM and ELKRAFT have a built-in interest and knowledge, which leads towards a systematic support for a centralised power plant development on the basis of fossil fuels.

The system is conducive to increased consumption because (1) the internal organisational dynamics causes the current motivation to construct new power stations. In ELSAM's area this is by means of competition between the seven power companies and, in ELKRAFT's area, this is by means of employment interests within the organisation. (2) The electricity distribution companies which urge consumers to economise on consumption pay the same to new power stations per kWh as those who encourage increased consumption. (3) The electricity consumer who economises on consumption pays the same to new power stations per kWh as those who do not economise. (4) Democratic influence increases with increased consumption.

The direct electricity system is relatively[20] dirigible because of (1) the independence of a horizontal and vertical ownership root-net, and (2) the lack of strong economic links to the economy of the public sector.

IV. The Links Between the Direct and Indirect Electricity Systems and the Democratic Process

When analysing the electricity system, it is not sufficient to look at the direct electricity system. One must also look at the relationship between the direct electricity system, the indirect electricity system, the consumers, and the Government.

When analysing these questions, the following statements can be drawn.

(1) There is no vertical integration with fuel procurement. This differs from many other European electricity systems. In the German case, the lignite mines are owned by the same organisation which owns the power plants as well as the transmission and distribution system. No coal miner will demonstrate if the use of coal decreases!

(2) There is no great degree of horizontal integration in the subcontractors for technology for power generation. In the German electricity system, for example, where the largest German company (RWE

---

[20] Relatively in relation to the power system in, for instance, Germany.

Energie) is a part of a group which also includes building contractors who construct power and smoke purification plants.

(3) There is no horizontal integration with other fuel companies. This is also different to the German system, where the VEBA Group, which owns PreussenElectra, is also the owner of a company which sells fuel oil. In the VEBA case, the establishment of co-generation units would decrease the turnover in the VEBA subsidiary VEBA Oil Company. The interest against co-generation is therefore much stronger in Germany than in Denmark.

(4) There is no horizontal integration with other large industries. This is different from the VEBA Group, where PreussenElectra has sister companies dealing with production of chemicals, transportation, oil, etc. Therefore there are no Danish industries which are interested in high electricity prices, whereas in the VEBA Group, members of this group will sometimes gain more than they lose by high electricity prices.

(5) At the same time there will be no specific low electricity prices for large consumers, such as those enjoyed by large firms in Germany.

(6) The Danish electricity is consumer-owned, and the profit remains with the electricity consumers in the form of lower electricity prices, whereas in the German electricity system, the profit goes to the shareholders. In the case of PreussenElectra, these are private shareholders, and in the case of RWE, these are regional and district authorities.

(7) In the Danish consumer-owned system, there are no strong interests working for higher electricity prices, whereas stockholders in a joint-stock company have a motivation for higher electricity prices.

(8) There is no interweaving of a specific electricity supply system's revenue and the public sector's taxation revenue. The public sector, with regard to its revenue, is independent of the electricity system's actual structure. This should be seen in relation to the Danish natural-gas companies, where the local authorities are financial guarantors and thus economically dependent on the sale of natural gas. In Germany, local authorities are dependent on concession levies which the electricity companies pay to them. Thus the public sector, with regard to their revenues, are dependent on the share dividends of the electricity companies. A reduction in electricity consumption or the establishment of independent producers will, in the German case, result in reduced revenues for regional and district authorities. In the Danish system, such a dependency does not exist.

V. Public Regulation and Development

The regulation and development problem on the energy scene is, under the present conditions, that the new technological solutions require changes at the level of techniques, knowledge and organisation. Under such conditions a regulation model along the lines of negotiated regulation has a fundamental problem, which is that the new techniques, knowledge and organisations are not present in the negotiation situation. This, naturally, has as one of its results that the necessary new technological solutions do not receive acceptable treatment in negotiations. We know, of course, that wealthy and focused[21] minority interests, like power companies and natural gas companies, have the will and resources to lobby for their interests with success. However, we also wish to emphasise that a functioning democratic process is the only place where there is always a potential majority against any short-term and narrow economical-interest group. The democratic process is a potential forum for radical technological innovations which are often against the inherent interests of the existing dominant firms in the market.

Until the mid-1970s, there was very little public regulation of the electricity system. The Electricity Supply Statute (1977) changed this, and since then there has been growing public attention and 'intervention in the affairs of the electricity system'.

The main questions at the end of the 1970s and in the early 1980s were nuclear power and the introduction of natural gas. Nuclear power was removed from the political agenda in 1984, whilst natural gas represented a new fuel-distribution infrastructure, which formed the base for a new competitive situation for the Danish electricity system. The anti-nuclear power movement was very strong, and educated a generation of people interested in energy, who are now working at universities, in consultancy firms, in the Ministry of Environment and Energy, in the school system, etc. This movement has continued as a sociological and psychological environment behind the energy-planning discussion in the 1980s and 1990s. Research funds were distributed to the development of prototypes of renewable energy technology, etc. from an institution which was totally independent of the old energy companies (the 'Styregruppen for vedvarende energi').

When this type of legislation was possible, it was linked to the fact that there are many parties in the Danish Parliament. The old, large parties (the Social Democrats, the Conservatives, and the Liberals) are all linked to interests connected to the fossil-fuel companies. Some of the small parties, however, are relatively independent of these links, and at the same time they sometimes have influence. For instance, in

---

[21] By 'focused', we mean that power companies are mainly producing power. The alternatives to power companies are often electricity conservation measures, which have to be taken in companies having totally different main purposes.

1989, the Minister of Energy was a member of the small Liberal Radical party. He changed the energy policy and was responsible for the new 'Energy 2000' plan from 1990.

The end of the 1980s witnessed the completion of the natural gas network. This network represents a physical infrastructure which makes possible an extensive development of the co-generation of power and heat, on the basis of natural gas, everywhere in Denmark.

The 1990s is a period in which the Danish electricity system has been and will be confronted with problems of hitherto unseen dimensions; at the same time, they will be of a character which the electricity system is unaccustomed to dealing with, and perhaps unable to solve. The most important external causes of these problems are environmental concerns, and increased international competition for electricity consumers, in connection with the single market.

The public regulation process has been (and still is) a dialectic process, on the one hand between 'negotiated regulation' between the central administration and the traditional power companies, and on the other hand active and open public debate with participation from grass-roots movements, proponents for the new technologies, the general public, and Members of Parliament. This dialectic process has sometimes resulted in victories for the power companies, and sometimes in victories for grass-roots movements and public debate.

If we start by examining the environmental problems, we can see that they were the cause of a massive Parliamentary majority supporting the goal that Denmark, in the period from 1988 to 2005, must reduce CO_2 emissions by at least 20%. This goal has been sought since 1990 by means of the Ministry of Energy's energy plan 'Energy 2000'. The present Ministry of Environment and Energy is continuing along the lines of 'Energy 2000', as it is illustrated in 'Danmarks Energifremtider' 1996. With regard to the supply of electricity, this plan calculates that (1) the installed capacity of decentralised co-generation of power and heat must be increased from the present 1400 MW to 1700 MW in the year 2000, and to 2000 MW in the year 2015. This should be seen in relation to a total installed effect of 8000 MW. By the year 2000, approximately 50% of Danish heating will be covered by supply from co-generation plants. (2) The effect from wind generators must be increased from 550 MW in 1996 to 1500 MW by the year 2005. The consequences of this are (although not clearly expressed in 'Energy 21') that no new coal-fired generating capacity will be built for a long time. With regard to the consumption of electricity, it is calculated that the total consumption of electricity will increase by about 12% in the period up to the year 2005, and thereafter will remain at that level.

Recently, Parliament's demands of the electricity system have been specified, so that the system has been charged with ensuring that CO_2 emissions are reduced by at least 20% of the 1988 level by the end of the year 2005. The politicians have not only been talking about green plans. They have also backed these plans by actual measures for their energy policies. Amongst these measures are the following.

(1) The introduction of a system of energy levies, including a CO_2 levy. This levy is 0.65 Dkr per kWh for households and a number of service enterprises, amounting to about 60% of the kWh price. The levy for industry, agriculture, market gardens, etc., is only 0.10 Dkr per kWh, amounting to 15% of the kWh price.

(2) A CO_2 subsidy of 0.10 Dkr per kWh for electricity produced by co-generation plants which are powered by natural gas.

(3) In 1994, the Integrated Resource Planning Statute was brought into operation. This Statute requires the electricity system to prepare 20-year energy plans every second year. On this basis, citizens will be kept aware of whether investments are to be made in plants, or whether consumers are to be affected by investment in energy effectiveness.

(4) In December 1995 a law was passed by Parliament establishing a set of regulations regarding the sale of electricity from decentralised co-generation plants. The law states the right to sell electricity from decentralised co-generation plants to the public net, at a price equivalent to the long-term avoided costs in the electricity system. This means a payment, including the capital costs per kWh of large, coal-fired power plants and the transmission network.

(5) On April 16, 1996, a legal proposal regarding a certain 'liberalisation' of the Danish electricity market was made by the Ministry of Environment and Energy.[22] The main content of this law is that: (a) Electricity distribution companies and companies with a consumption above 100 GWh/year are allowed to buy electricity in the marketplace, for instance from Norwegian or Swedish power companies. (b) Electricity from co-generation plants and renewable energy plants has priority in the market. This means that any distribution company is obliged to buy a proportion of the total electricity production from these plants. The production from these plants will often be 80–90% of the whole electricity market in Denmark, and this market share is increasing. During the summer months, this proportion is naturally much lower. (c) If any costs are connected to the above-mentioned obligation, these costs are equally distributed to the consumers of electricity. This law was introduced in order to protect the Danish environmental 'public service' policy at the same time as the market was partly opened

---

[22] Forslag til Lov om ændring af lov om elforsyning, Miljø og Energiministeriet 16/4, (1996).

for the new market regime. The new 'Energy 21' maintains as its policy a continued support of consumer ownership in combination with a non-profit obligation.

As for the public regulation process in Denmark, the following can be concluded. (1) It has, in several cases, been possible to mobilise the democratic process in such a way that radical technological changes have been introduced and implemented. This is, among other things, due to a combination of: (a) many parties in the Parliament, some of which have no links or interests directly linked to the power companies, (b) easy access to communication with members of Parliament, (c) grass-roots movements working during the 1970s and 1980s educating a lot of people within the energy area, and in that way spreading this knowledge to almost any level in society. (2) The financial possibility of independent research at universities and new centres for renewable energy.[23]

The latest legislative move regarding market access, and a continuation of the energy policy from 'Energy 2000', shows a beginning change from 'challenged negotiated regulation' to 'challenged legislative regulation'. This has been made necessary, among other reasons, because of the pressure of the changing market regimes around Denmark.

VI. The Dynamics of the Danish Electricity Service Supply System and the Public Regulation Process

Any description of a (electricity) system limits the mind to specific questions, levels of aggregation and areas of focus, and therefore also to specific reform proposals. In a situation where radical technological changes are needed, it is extremely important not to fall back upon our traditional way of describing an electricity system. One must design the description in accordance with the purposes of the analysis, and be conscious of the limitations built into the specific way of description. We describe the structure and dynamics of the electricity service supply system in accordance with the purposes of our analysis, which is to evaluate the links between energy effectiveness and the political economy of electricity systems. When analysing the electricity system with these purposes in mind, we find it necessary to perceive the totality of direct and indirect electricity service supply systems, together with the democratic and public regulation process. Our analytic framework is described in Fig. V.3.

The main features of the analytical framework in Fig. V.3 are the

---

[23] For example, the 'Peoples Centre for Renewable Energy' in Ydby.

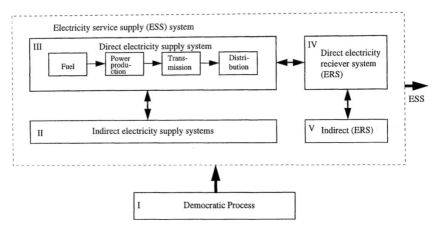

Fig. V.3. The electricity service supply system and the democratic process.

following. The concept of electricity systems as service supply systems producing services such as cooling, heating, motion, etc., by means of a certain amount of kWh electricity production and by means of investments in electricity receiver systems. When analysing political dirigibility, it is necessary to divide direct and indirect supply and receiver systems, as the indirect electricity systems have important channels of political influence. For the same reason, it is important to look at the electricity system as a part of a total system, including democratic processes.

At the beginning of this chapter, three main questions were asked. These will now be answered in the following.

VI.A. Question 1. Which type of dynamics is built into the organisation of the Danish direct electricity supply system?

VI.A.a. Is the Danish electricity supply system cost effective?
The Danish electricity supply system is cost effective with regard to its centralised power generation on the basis of fossil fuels. The Danish electricity system is interesting as an organisation, because it has been able to establish a cost effectiveness for systems which are based on fossil fuels and centralised power generation. This cost effectiveness, within a given paradigm, is implemented in spite of it being a monopolistic system. The cost effectiveness appears to be caused by the system practising consumer profit-sharing, openness in electricity prices and electricity generating costs, in combination with co-operative ownership

organisation, where the consumer representatives, in spite of low election participation, have relatively effectively represented a desire for cost awareness, especially at the level of the electricity distribution companies.[24]

VI.A.b. Is the Danish electricity system conducive to increased consumption?
The Danish electricity system is conducive to increased consumption, on the following grounds.

1. A capital transference system where the electricity distribution company (and the consumer) which initiates reduced consumption pays the same per kW new capacity as the electricity distribution company (and the consumer) which furthers increased consumption.
2. A tariff system where the average price per consumed kWh falls according to the size of the increased consumption. This is caused by a system of consumption-related levies which, in a number of areas, are as high as 600–800 Dkr per consumer household.
3. The EU's minimum electricity prices for the private sector. This is caused by the Danish electricity system not being liable for income or capital taxation.

VI.A.c. Is the Danish electricity system, of its own volition, and without Parliamentary intervention, able to implement the necessary radical technological changes?
The answer is that the Danish electricity supply system is technologically conservative, and that it is the centralised supply-oriented paradigm which needs to be answered. The Danish electricity system, of its own volition, does not take the initiative to introduce radical technological change, and it systematically opposes the implementation of such changes. Since the beginning of the study period (1972), the electricity supply system has not taken any initiative for a general introduction of electricity saving, decentral co-generation plants, or utilisation of renewable energy resources. The demand for the introduction of these technologies in energy policies has come, and still comes, from grass-roots movements, in combination with the parliamentary process. Reforms, which make it possible for these technologies to enter the market, have also come from popular movements and Parliament.

During the course of the whole period, the electricity supply system

---

[24] This is described in Hvelplund (1995) (to be published in October, 1996) in which it is described how labour productivity at the power-company level has fallen, but has risen at the distribution-company level.

has systematically opposed the introduction of these new technologies. The system's strategy has been:

1. to make it difficult for a competitive technology to enter the political agenda;
2. if the new technology has been put on the political agenda then the practical implementation has been opposed; and
3. if this opposition has not been successful, then it attempts to take possession of the new technology.

Technological conservatism appears to be caused by an electoral system with a series of indirect elections which are directed towards strategically designed organisations. ELSAM and ELKRAFT are dominated by representatives from the centralised power technologies.

VI.A.d. Can the Danish electricity system compete under the new international regulation regime?

Yes, due to cost efficiency and a high level of consolidation, the danger of price competition in the Danish home market is very low. The real danger is linked to the sale of parts of the system to foreign power companies, for example PreussenElectra. Due to the Danish consumer-ownership structure, it is possible to prevent foreign companies from buying the whole Danish power system.

VI.A.e. Is the Danish electricity system organised in such a way that it can be directed by the Parliamentary system?

This is relative, in comparison with the lack of political dirigibility which is found, for example, in the German and French electricity systems. Amongst the causes of this relative political dirigibility are:

1. it is a self-financing, non-profit system which can survive financially in a falling market;
2. there is an openness on prices, costs and development plans;
3. the Danish direct electricity supply system is a minor part of the whole electricity supply system; and
4. the public accounts are not dependent on income or expenditure which is related to any specific electricity supply structure.

However, this dirigibility is not worth very much if the electricity system captures the central administration, and thus secures the steering. In recent times, there has been a tendency in that direction through the establishment of a number of closed committees consisting of members from the electricity sector and the central administration. One of these committees, the Electricity Strategy Committee, has been directly involved in the preparation of the new Electricity Supply Statute.

VI.A.f. Is the Danish political system able to regulate the power companies?
The Danish political system has been able to introduce a set of regulation measures establishing radical technological changes, changes which often have been against the will of the power companies. The process has been a dialectical process between a tradition of negotiated regulation and another tradition of public interest and interference.

The conclusion is that, with regard to the dynamics of the direct Danish electricity service supply system, it is a cost-effective, consumption-encouraging, technologically conservative and relatively politically dirigible system. The technological conservatism and the strong encouragement to consume are not in harmony with the demands for resource saving and the capability for technological change. The democratic process often encounters many difficulties, and the transaction costs of regulating the utilities with their present organisation seem too high.

VI.B. Question 2. Which dynamics are built into the Danish democratic system regarding its ability to regulate the Danish electricity system?

The Danish Parliament has so far shown some capability regarding the introduction of radical technological changes on the energy scene. This capability arises from a combination of the development of a rather strong public debate on energy questions in the 1970s regarding nuclear energy, and a Parliament with small political parties, which are independent of the 'lobbyism' of the power companies. The public debate was developed by grass-roots movements which again educated a lot of people in energy questions. So far, the regulation model has been what we call 'challenged negotiated regulation'. This model of regulation no longer seems to be sufficient, because of the new international development and the need for the development of decentralised technologies.

VI.C. Question 3. Which dynamics are built into the combination of the electricity system and the democratic system regarding its ability to establish regulation procedures between the new European market regimes and the environmental 'public service' goals?

The latest development has shown tendencies in the direction of regulation by (negotiated) legislation instead of regulation by negotiation. A law stating the conditions regarding the sale of electricity from auto-producers to the public net was accepted in Parliament at the end of 1995. According to the law, the power companies are obliged to pay a price for electricity from co-generation units equivalent to the long-term marginal electricity production costs. At the time of writing (spring 1996) a law regarding third-party access was introduced by the government, stating the rules regarding third-party access in the Danish electricity system. We call this 'negotiated legislation', as the existing power companies have been closely linked to large parts of the legislation process. The exception is the law regarding the conditions of sale of electricity from co-generation plants, which was not negotiated with the power companies, and was introduced against their will.

The necessary technological changes require more initiative by new organisations. It is not possible for the old power companies to develop and introduce these new technologies. New dynamic organisations and organisational models are needed. Regulation, at this stage of the technological development, has to introduce such organisational measures, but so far this has not been done. Whether Parliament will be able to introduce such regulation measures is the big question for the time being.

VII. Proposals for Changes in Public Regulation and the Electricity Service Supply System

The situation has changed. We have a new development on the international scene and we are at a stage of the technological development where a new public regulation paradigm is needed. Together with a detailed knowledge of the existing organisational setting, this has to be taken into consideration when the public regulation strategies of the future are designed.

VII.A. The dynamics of the role of the Danish electricity supply system

The developments of the last two decades have shown that the Danish electricity service supply system is technologically conservative, in as much as it has an inherent need for expansion in the direction of development towards large central power plants, in combination with major expansion of the high tension grids.

The system's employees do not possess the competence to take the initiative to implement technological changes, which would require organisational change in the electricity system. This can be illustrated by the fact that, in the 20 years of the study period, there are no examples of any of the 11,600 employees in the electricity system having opposed the fundamental energy policy attitudes of the system's management. Initiatives for the introduction of wind generators, decentralised co-generation plants, industrial co-generation and electricity savings have all come from outside the electricity system. These initiatives have all been met with opposition from all parts of the electricity system. However, the opposition has not always been so severe that some of the initiatives could not be implemented. This is caused by a number of conditions in the Danish parliamentary system, and in the Danish electricity system's historical organisation, with consumer ownership, non-profit[25] principles and relative openness on prices and costs. As opposed to the natural gas system, there is no significant confusion between the economy of the electricity supply system and the economy of the public sector, which serves to increase the electricity system's political dirigibility. These very positive aspects of the Danish electricity system should be retained, in connection with the implementation of future reforms for the electricity area.

The present development in the electricity system's organisation is characterised by the inherent need for expansion, as is being experienced in the German, Swedish and French electricity systems, and in EU activities towards a change of regulation on the European energy market.

If one is to understand the current pattern of reaction in the Danish electricity system, it must be comprehended as a battle on two fronts. Firstly,[26] opposition continues against wind generators, decentralised co-generation plants, industrial co-generation and electricity savings on the home market. This opposition swings continually between opposition and, where this does not succeed, takeover. Secondly, the threats and possibilities from relations with PreussenElectra, Vattenfall, Electricité de France and EU regulation changes need to be managed.

The electricity system's desire for an expansion of the capacity of the power plants and high-tension grids must be seen as a stage in the participation of the Danish electricity system in the struggle/co-operation with large neighbouring electricity companies, on their technological premises. From the perspective of the power companies, it is

---

[25] It is important to indicate that this is a consumer-profit system, in which cost awareness is to the benefit of the consumers by means of the return of profits to the consumers in the form of lower prices.

[26] It is not to be expected that the electricity system is incited to major implementation in any of these fields. It is difficult to believe in an initiative where electricity savings are of such a volume as to make new generating capacity unnecessary for the next ten years. Of course there are groups within the electricity system which are working seriously and determinedly with electricity saving, etc. However, these initiatives have not shown themselves to be powerful enough, and the Danish Association of Power Generating Companies has continued to oppose determinedly measures to convert electrical heating, in 1994 and 1995.

important that the integrated resource planning does not prevent the construction of a planned 400 MW power plant, Avedøre II, which is a part of the co-operation with the large Swedish power company, Vattenfal, or the extension of the KONTEK connection, which is part of the agreement with the German VEAG/PreussenElectra/RWE, etc. electricity companies. A takeover strategy, or a practice whereby the electricity system (with the consumers' money) gains a monopoly on electricity conservation activities has been established. In the closed Electricity Strategy Committee, the implementation of the Statute on Integrated Resource Planning has been limited to the electricity system (with the consumers' money) implementing electricity-saving activity, concurrent with an increase (or at least no reduction) in electricity consumption.[27] Thereby, electricity conservation is controlled in such a way that the Danish electricity system's interest in the construction of new power plants and participation in the struggle/co-operation with large neighbouring power companies, is strengthened.

At this time, the confrontation on two fronts appears to be the dominating interest within the electricity system, but it is not the current official interest of energy policy. Confrontation strategies have meant that public regulation and the democratic process are becoming more closed with regard to the new fuel-economising technologies, and do not lead to solutions which are in accordance with the aim to reduce CO_2 emissions by 50% before the year 2030.

VII.B. The environmentally necessary technologies

At the same time, we have come to a point in technological development in the energy scene in Denmark where it is no longer possible to gain much by using 'end-of-pipe' solutions. It is no longer possible just to build another, better power plant, or to clean the smoke. The new technologies are decentralised, and have to be developed, introduced and implemented by organisations which often do not have energy as their main area of interest. At this stage of technological development, new infrastructures, organisations, calculation methods, etc. are needed.

---

[27] In connection with the argument for the new Avedøre II generating station, ELKRAFT assumes that electricity consumption will increase (unless there is public intervention) by 12% over the next seven years, even though the increase in the last seven years has only been 3–4%. If the results of electricity saving, supported by energy policies, are deducted from this 12% increase, then the result is a stagnation in electricity consumption.

VII.C. The international development

New market regimes are at present being introduced in Europe. The actors in the Danish energy scene are increasingly dependent on what is happening in Scandinavia and the European Union. Regulation by negotiation may be increasingly difficult with internationalisation and the increasing difficulties in identifying with whom the administration should negotiate. It is in this landscape of changing technological and international conditions that the following proposals should be seen.

VII.C.a. Proposal to reduce the technological conservatism
Internally in the electricity system. Technological conservatism appears to be the result of the electoral system, with a series of indirect elections which are directed towards strategically designed organisations. ELSAM and ELKRAFT are dominated by representatives from the centralised power generation technologies. The negative effects must be changed, by means of (1) changing the election procedures of the electricity system to more direct elections, and (2) strengthening of those parts of the electricity system which are independent, and also strengthening of the opportunity for the public to participate in the development of alternatives to the proposals of the electricity technocracy.

In relation to the public administration and political dirigibility. (1) Changing the capital transference mechanisms in order that they may be utilised by groups which are independent of the present electricity organisation to construct power plants which are able to economise in electricity and in the use of fossil fuels. (2) Establishing new corporation committees in the energy administration.

VII.C.b. Proposal to maintain the cost effectiveness
The cost effectiveness appears to be caused by a system with the following characteristics which should be maintained: (1) a consumer-profit system, (2) openness on electricity prices and electricity generating costs, and (3) the co-operative ownership organisation, where the consumer representatives, in spite of low election participation, have relatively effectively represented a desire for cost awareness, especially at the level of the electricity distribution companies. These characteristics of the electricity system must be maintained in order to sustain the cost efficiency and the price structure of the Danish electricity system.

VII.C.c. Proposal for change from consumption-furthering to consumption-reducing dynamics
Changes in the direction of progressive electricity prices and capital transference regulations which support electricity saving.

VII.C.d. Proposals for the strengthening of the democratic process
(1) More public access at all levels, (2) that committees in the central administration have members not only from the old power companies but also from green organisations, the new renewable energy and conservation technologies, etc. The tradition of negotiation in committees with members from the old power plant companies should therefore be changed, and (3) that the new law on integrated resource planning should be changed in such a way that groups other than the old power plant companies have access to independent planning resources and investment funds.

Literature

Danmarks energifremtider (1996) Miljø- og Energiministeriet 1996.
Hvelplund, Frede (1996) *Det danske elsystems politiske økonomi* (to be published in October, 1996)
Lund, Henrik and Hvelplund, Frede (1994) *Offentlig regulering og teknologisk kursskift*, Aalborg Universitetsforlag, 1994.
Ministry of Energy (1990) *Energi 2000*.
Ministry of Environment and Energy (1996) *Energi 21*.

Chapter VI
The Dutch Electricity Reform: Reorganisation by Negotiation[1]

M. J. ARENTSEN, R. W. KÜNNEKE AND H. C. MOLL

I. Introduction

In The Netherlands, the electricity market is in transition from small-scale local systems to a highly regulated public monopoly to (possibly) a liberal market system. The process started in the mid-1980s and is still continuing. The system is in upheaval, and the outcomes are not yet clear. Producers have merged together in an attempt to realise economies of scale and to become competitive in the (expected) international electricity market. Distributors have found new opportunities to penetrate the production sector, attacking the old and stable process of planned electricity production. Private firms invest in auto-production facilities, in search of lower electricity prices. In electricity distribution, an unexpected concentration process has occurred which has significantly reduced the number of firms in this sector. This process has developed without any clear end being defined by the sector or the Dutch government. However, very recently the Dutch Ministry of Economic Affairs proposed a new sector organisation which might serve as a blueprint for new and stable institutional arrangements.[2]

Until now, negotiation between the sector and central government has been the dominant regulation style in the Dutch electricity market reform. The reform is managed by mutual agreements between the sector and the government which have resulted in voluntary policy changes of distributors and producers, in some cases carefully pushed

---

[1] The authors would like to thank Professor Dr G. W. Johnson (Department of Political Science, Auburn University, USA) for his valuable comments on an earlier version of this paper.

[2] 'Derde Energienota', 1996 (Third Policy Note on Energy, 1996).

by legal enforcement. This process, based on depolitisation, is a quite typical Dutch practice of handling conflicts and institutional change.[3]

Due to the changing perceptions of politicians about the public tasks of electric utilities, in combination with the changing regulatory role of the public authorities in this industry, the Dutch government initiated reform. The reform process was formally initiated by proposals for new legislation in the 1980s. The announcement of new electricity acts for the organisation of electricity production and distribution forced the sector into a transformation process, which, according to the sector, should be handled by self-regulation.

Before 1985, the electricity system consisted of many relatively small production and distribution companies, each serving specific segments of the electricity market. The transformation process was initiated in the period between 1985 and 1989. During these years, the electricity system was confronted with new dynamics, challenging the existing market structures and settled positions. In 1989, a new Dutch Electricity Act, which regulates mainly the production sector, came into force. This act facilitated new opportunities for market parties. Some of them took these opportunities, such as the distributors, whereas others had to defend their positions, such as the producers of electricity. The Electricity Act of 1989 initiated a process of increasing orientation toward more flexible and liberal market relations. The reform process until now has resulted in an unclear and unstable market structure. The market is not yet in an institutional equilibrium, as shall be discussed in detail below. Several different proposals for a liberal market structure have been developed, meant to be 'the new steady state' of the Dutch electricity market.[4] This process resulted in 1996 in the above-mentioned policy note of the Dutch government of a new market-oriented sector organisation. As a separate political activity, a distribution act prepared

---

[3] In 1968 the Dutch political scientist Lijphart published a book about this phenomenon, entitled 'Verzuiling, pacificatie en kentering in de Nederlandse politiek' (Denominationalism, pacification and turn of the Dutch policy). The central thesis of the book is the idea of pacification, a mechanism to govern societies with manifest societal cleavages, on socio-economic structure, religion, ethnography and the like. Until the mid-1970s, the Dutch political system was dominated by coalitions of the religious block, the liberal block and the social democratic block. These blocks had strong ties with societal groups, based on socio-economic and religious cleavages. To handle these societal cleavages, the Dutch developed the habit of consensus-building at the top. This practice was quite typical until the mid-1970s, but decision-making by consensus-building is still quite common in The Netherlands.

[4] Examples are: Horizon 2000 (1996), *Visie op de ontwikkeling van een klantgerichte electriciteitsmarkt in een Europees perspectief*, Andersen Consulting, The Hague, 1993; *Rapport van de groep van Negen* (1994), Arnhem; *Verzekeren van een passende elektriciteitsvoorziening voor de toekomst* (1994), McKinsey & Company, Amsterdam.

by the government has been discussed for about a decade, and was very recently introduced in (both chambers of) Parliament.

This chapter is organised as follows. Section II presents some actual features of the recent Dutch electricity system up to 1995. Section III evaluates the reform process since the mid-1980s. The elaboration concentrates on institutional, regulatory and technological changes of the system. Section IV discusses the results of the reform with respect to criteria of institutional stability, economic performance, public tasks and innovativeness. The chapter concludes with some prospective reflections.

II. The Dutch Electricity System in the Mid-1990s

II.A. The institutional framework

Fig. VI.1 illustrates the general institutional framework of the Dutch electricity sector up to 1995. Four regionally based companies, united

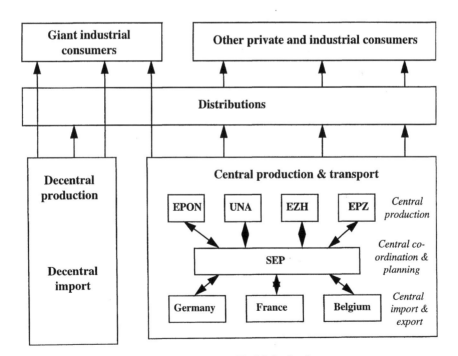

Fig. VI.1. Structure of the electricity sector in The Netherlands.

in the SEP,[5] produce and transport electricity in The Netherlands. SEP is responsible for the import and export of electricity at the national level. There is a separate, vertically disintegrated distribution sector.

In 1993 there were some 29 distribution companies, most of them not only distributing electricity, but also natural gas for residential heating, district heating and central cable services for radio and television. In 1993 the distributors delivered electricity to some 6.7 million customers. Households took some 18,000 GWh, and industry some 26,000 GWh that year.

Some consumed electricity is produced or imported decentrally, as illustrated in Fig. VI.1. The Electricity Act of 1989 allows for some decentralised production and import of electricity. Giant industrial consumers have free access to international electricity markets; however, none of them exercised this right until recently, because they are charged quite extensively for using the grid. Distributors are allowed to produce electricity in plants smaller than 25 MW of installed power. Industrial companies have the legal possibility to produce electricity without limitation of the size of their generation capacity. Distributors have the obligation to buy this surplus electricity at a fixed price, which is cost-effective. As will be illustrated below in more detail, this decentralised production capacity challenged the national production planning of the four giant producers united in the SEP, enhancing over-capacity at the national level.

II.B. The actual functioning of the Dutch electricity system up to 1995

At the national production level, SEP is responsible for the co-ordination of electricity production on a daily basis, and for the contracts for the import and export of electricity. To secure a long-term supply of electricity (investments in new power plants and the closure of the old ones), the production companies develop joint national plans every two years, which have to be approved by the Ministry of Economic Affairs. This 'Electriciteitsplan' is binding for the four producers and SEP.

The co-ordination of central electricity production is very strictly arranged. Based on a production-cost optimisation program, SEP orders the producers to generate electricity in the indicated plants. Regional surpluses of electricity are passed over by SEP to producers with a shortage to serve their region. The benefits of this procedure are distributed among the producers with the cost-pooling system. This system

---

[5] This abbreviation stands for the **Samenwerkende Electriciteits Producenten** (Dutch Electricity Generating Board).

results in a uniformity of production costs. In the same way, the benefits of import and export transactions by SEP are divided. Although only the distributors are legally excluded from import and export activities, SEP has a *de facto* monopoly in the international trade of electrical power.

Distributors buy electricity from their regional producer, for the national pooling price, plus costs of transport and delivery of the regional producer. Legally, distributors have the possibility to buy electricity from a producer of their choice within the national boundaries of The Netherlands. This opportunity of 'horizontal shopping' is not effective, and thus not used, because of the absence of effective price differences.

Recently, several giant industrial consumers used the legal opportunities of electricity production not only for supporting their own consumption, but also for considerable contracts with local distributors. Several distributors established joint ventures with industry in order to build decentral capacity, which in all cases is CHP. This decentral electricity production gained a significant market share of about 22% of national electricity consumption. This success of decentral production untied the strict relations between producers and distributors in the Dutch electricity system. Although the amount of decentralized produced electricity is still limited, the decentralized production activities have become the first serious challenge to the production monopoly of the SEP. Recently, CHP has received substantial societal (and public financial) support for environmental reasons, tackling large-scale electricity production by SEP, in favour of the CHP investments of the distributors.

II.C. Some basic statistics of the Dutch electricity industry

In this Section, some basic statistics of the Dutch electricity system are presented, highlighting some technical and economic details. The data are summarised in Table VI.1.

A very important characteristic of the Dutch electricity system is the dominance of natural gas as a primary energy source, with a share of roughly 55% of all fuels used in central and decentral production (see Table VI.1). This dominance is a consequence of the huge national gas reserves in the northern part of the country. The so-called Groningen field is considered to be one of the biggest in the world, and is very cheap to exploit. As will be discussed below, natural gas only recently became the dominant fuel to produce electricity. For years, coal dominated as the primary fuel for the generation of electricity.

In the period 1970–1995, the efficiency of gas-fired and coal-fired

Table VI.1. Key data of the electricity supply in The Netherlands

	1984	1989	1990	1991	1992	1993
Number of generating companies	16	5	4	4	4	4
Number of distribution companies	69	48	43	41	34	33
Staff	26,207	25,854	24,975	24,390	23,890	23,525
Investments (in million NLG)	1675	2090	2050	2685	2905	3090
Turnover (in million NLG)	10,673	8,867	9281	9521	9545	9940
Installed generating capacity in MW (under SEP)	14,890	15,689	15,142	14,867	14,602	14,458
Maximum load (in MW)	9470	10,400	10,760	10,860	10,620	10,910
Net output (in GWh) (under SEP)	53,006	60,059	57,524	59,040	60,260	58,760
Primary fuel (%)	27.3	37.4	45.0	38.9	38.0	37.3
Coal	0.9	0.6	0.3	0.7	0.2	0.3
Oil	63.8	54.2	47.6	53.8	54.4	54.6
Gas	8.0	7.8	7.1	6.6	7.4	7.8
Uranium						
Average purchase price of all fuels (in NLG/GJ)	11.54	5.26	5.45	5.56	5.3	5.25
Thermal efficiency (%) (fossil fuels)	39.2	40.4	40.4	40.6	40.9	40.7
Length of high voltage grid (in 1000 km)	91.0	99.4	99.8	100.5	100.9	101.3
Length of low-voltage grid (in 1000 km)	118	127.7	128.6	129.5	130.3	131.0
Imports (in GWh)	3649	5289	9677	9778	8905	10,565
Exports (in GWh)	0	339	471	624	228	55
Number of customers (in 1000)	5747	6309	6389	6497	6595	6666
Consumption (in GWh)	54,970	64,117	66,502	68,635	69,895	70,740
Average price for households (in cents per kWh)	23.8	17.7	18	18.1	17.8	18.2
Average price for industry (in cents per kWh)	12.9	9.6	10.1	10	9.8	10.1

Source: SEP (1994), *Electriciteit in Nederland, 1993,* Arnhem.

electricity plants increased substantially. For new gas-fired plants efficiency has risen from about 40% to more than 50% (the newest plant becoming available in 1995/1996 has a designed efficiency of almost 55%). The integration of a high-temperature gas turbine and a steam turbine (a so-called STEG turbine) results in a more efficient utilisation of the fuel energy than in a conventional turbine. The conventional coal plant, based on pulverised coal combustion and tail-gas desulphurisation, has an efficiency of 35–38%. New coal technology, based on an integration of a coal gasification process and coal gas combustion in an STEG turbine, results in an overall efficiency of above 40% and an effective prevention of SO_2 emission.

The significance of nuclear power in generating electricity has always been very limited in The Netherlands. Although the Dutch participated in the nuclear programme 'Atoms for Peace', initiated by the Americans just after World War II, a mature nuclear industry never came into existence, as happened in France or Belgium. In contrast to the Belgians, the Dutch had no national incentives to develop a nuclear industry because they had their own reserves of coal and natural gas and no direct access to uranium, as the Belgians had in their former colony of Congo.

The Dutch erected two nuclear power plants for national technological and economic reasons. In the mid-1980s, the Dutch government decided on the extension of nuclear energy for national electricity production and faced extensive societal resistance. To legitimise pro-nuclear decision-making, the government organised a societal debate on energy-supply politics. Just before the pro-nuclear decisions had to be taken, the world was facing the Chernobyl accident. The impact of that accident, everywhere in Europe, and especially in The Netherlands, diminished the remaining legitimacy of extending nuclear power. Thus the Dutch government withdrew the proposals on new nuclear power plants. Since then, nuclear power has no longer been a serious technological option for electricity production. Last year the Dutch Parliament decided to close down the existing nuclear power plants by the year 2004.

Next we turn to the question of the actual functioning of the institutional framework. Looking at the other features summarised in Table VI.1, the following comments can be made. As mentioned earlier, the number of distributors and producers has declined drastically. This partly reflects the fundamental institutional change which took place in these years. This process will be addressed in depth in the next Section.

There has been a significant decline of staff in this sector, from (roughly) 26,200 to 23,500 persons employed. Meanwhile, investments nearly doubled, from HFL1.7 billion in 1984 to HFL3.1 billion in 1993. Turnover declined after 1984, but is increasing back to its former level. The installed capacity of the four central electricity producers has declined since 1989, as a consequence of the increasing importance of decentral producers. The net output of central production is relatively stable, at a level of 58,000 GWh per year.

The average purchase price of primary energy has been relatively stable since 1989, with a price of roughly 5.5 NGL/GJ. There are still new investments in high- and low-voltage grids, but both systems pro-

vide a good and sufficient infrastructure for the transport and distribution of electricity to all end consumers in The Netherlands. Hence the
number of customers is relatively stable.

The import share of electricity is substantial: about 15% of the electricity consumption is imported. The import of electricity nearly tripled in
the years between 1984 and 1993. The Dutch took advantage of cheap
electricity in France and Belgium (nuclear and hydro-power). German
electricity, mainly purchased from coal-fired generators, is imported to
meet the public task of fuel diversification.

The average price for households and industry has been stable since
1990, and compared to other European countries, is at a very low level.

III. The Emergence of the System Reform

The emergence of the electricity system as presented in Section II is the
result of the changing regulation of the Dutch electric utility sector. The
process started from a context of stable, small-scale local organisation
of production and distribution before 1985, went on to a period of
transition between the years 1985 and 1989, to continue until the
mid-1990s in a process of upheaval and uncertainty. Before going into
the details of the reform process, some of the most visible outcomes of
the reform are illustrated in Table VI.2, which gives an overview of the
distribution companies and the number of customers served.

In the period 1977–1984, the number of companies remained almost
stable. The total number of companies decreased by three due to the
merging of some very small-sized companies. The number of customers
served increased substantially in this period, so the other companies
increased in size proportionally. As shall be argued more in detail
below, four time periods can be distinguished in the development of
the reform of the Dutch electricity market.

In the period 1985–1989, the total number of companies decreased
by 30, mainly because of mergers or takeovers by large companies of
small-scale companies (less than 30,000 customers). Most of the newly
formed companies remained moderate in size (less than 60,000 customers). Only two relatively large new companies were founded in this
period, one with about 75,000 customers, and one with 215,000 customers. A majority of the companies in 1984 having more than 30,000 customers showed a natural growth of customers, sometimes accelerated
by the takeover of small companies; two of these disappeared by a
fusion of companies.

The period 1990–1995 is characterised by the takeover of small- and
medium-sized companies by big ones, and by the creation of three

Table VI.2. Number and size of electricity distribution companies and the number of customers served, 1977–1993

Number of electricity distribution companies by class Market share (% of total customers served)	1977	1984	1989	1993
Less than 10,000 customers	26	19	3	—
Market share (%)	4	2	0	0
Between 10,000 and 30,000 customers	29	30	14	3
Market share (%)	10	9	5	1
Between 30,000 and 60,000 customers	11	12	12	9
Market share (%)	10	9	7	5
Between 60,000 and 120,000 customers	6	7	10	4
Market share (%)	9	9	12	5
Between 120,000 and 240,000 customers	8	6	6	3
Market share (%)	28	19	18	9
More than 240,000 customers	5	8	9	10
Market share (%)	39	52	58	80
Total number of distribution companies	85	82	52	29
Total number of customers (thousands)	4742	5750	6309	6818

Not all data in Table VI.2 are consistent with Table VI.1, because of different data sources. References for Table VI.2 are: *Concentratie Nutsbedrijven* (1980), SDU, Den Haag; *Jaarlijkse Statistiek van de electriciteits voorziening in Nederland over de periode 1984–1990*, CBS, Den Haag; and K. Kort (1994), *Reorganisatie van deenergiedistributiesector in Nederland, Gas*, February, pp. 78–84.

mega-companies (above 750,000 customers) by the fusion of some large companies. The outcome of this process is that the ten biggest companies acquired about 80% of the market by 1993. This process of mergers and fusions continues.

In the period after 1996, a market-oriented system might emerge in The Netherlands, based on the latest policy proposal of the Dutch government. The details of the reform process will be dealt with separately for each phase.

III.A. Before 1985: small-scale stability

The Dutch electricity sector has its origins in the municipal producers/distributors which emerged during the turn of the century. Electricity production and distribution developed as a municipal task which later also became a responsibility of the provinces. The provision of electricity was considered a public task of common economic need, to be provided by public service departments of local or regional government (Simons, 1939). Though in principle electricity could be provided by private firms—actually common practice in the very beginning of electricity production and distribution in The Netherlands[6]—there were two important reasons for public provision. First, the infrastructure, i.e. the grid, makes use of the municipal road network. The municipal owner of this network is thus directly involved in the development and maintenance of the electrical grid. Secondly, the electric utilities are traditionally considered as natural monopolies. To avoid destruction of capital by means of inefficient double investments in the infrastructure, it was perceived to be socially desirable to have one grid, the use and access of which is regulated by public institutions. Regulation is necessary to protect consumers from the misuse of monopolistic power, in terms of extensive pricing and/or selective provision of this essential good.

Beside these primary regulatory goals, municipalities also used their ownership of public utilities for general socio-economic and financial goals. For example, the city of Amsterdam had a tariff structure which favoured small domestic consumers relative to industrial consumers, related to the distribution costs of these groups of consumers. In Rotterdam a very socially minded policy was introduced with respect to the possibility of disconnecting defaulters, which led to back payments of $60 million. For the municipal owners public utilities became an important source of income, especially in the late 1970s and early 1980s, in which period public budgets were cut. Single Dutch cities introduced a system of 'normalised profits' which had to be paid in advance to the municipality (Baake, 1988).

According to the growing technical need of co-operation between the local producers/distributors and the possible cost savings because of the emergence of economies of scale and scope, the number of electricity distribution companies declined quite drastically from the beginning of the century. In 1920 there were about 550 distribution companies, declining to about 200 in 1960, 100 in 1980, and 82 in 1985 (Brandsma, 1985). Of these 82 companies, 64 are horizontally integrated with gas and/or water and/or central antenna. During these years, all producers were integrated vertically with transport and distribution, resulting by the mid-1980s in some 14 main producers (Brandsma, 1985), who also became the biggest distributors. Ten of these firms are integrated vertically at the provincial level, the others are geographically related to the four largest cities in the urbanised western part of The Netherlands.

The first reconsideration of the role of municipalities in electricity production and distribution started in 1958. These reconsiderations initiated a process of concentration and vertical integration to increase the professionalism and the economic performance of the public utility

---

[6] The first electricity producer/distributor was a private firm in the city of Rotterdam in 1883

sector. In 1970, this process was picked up again by two advisory boards, which recommended concentration and vertical integration to the Dutch central government to increase the economic performance of the public utility sector.[7]

The start of the public debate on the restructuring of the public utility sector, which preceded the ongoing reform process, was actually initiated by the so-called 'CoCoNut' board.[8] This board, installed by the Minister for Economic Affairs, had to advise on the impact of concentrating public utilities. The CoCoNut board was strongly in favour of more concentration because it was the only tool to realise a significant cost efficiency in the public utility sector. The Dutch government took over the recommendations of the CoCoNut board and opened the dialogue with the utility sector on institutional restructuring. The utility sector, united in associations for electricity, gas and district heating, was willing to discuss the need for reconstruction, as far as the process could be governed by the sector itself, without the formal intervention of the central government. The sector underlined this position by installing its own advisory board (The Brandsma Commission), to make an inventory of the organisational alternatives for restructuring the sector in congruence with the official governmental goals. The Dutch government agreed upon the voluntary reconstruction process, but thought it wise to initiate new legislation to ensure that the voluntarily co-operating parties would take their task of reconstruction seriously. The Dutch government prepared a Distribution Act, to be used as a 'big stick' in case the self-regulating forces of the sector turned out to be unreliable or ineffective. The Brandsma Commission started working in 1985, marking the end of an area of institutionally stable and decentralised local utility structures.

III.B. 1985–1989: Transition

In 1985, the discussion on the structure and tasks of public utilities was opened by the installation of the Brandsma Commission. The participants concentrated on discussing several aspects of the future structure of public utilities. A very dominant aspect was the optimal scale of distribution companies. An extensive public discussion about the optimal scale of electric utilities led to different points of view. Central government opted for a minimum scale of 100,000 consumers, a number

---

[7] The Hupkes Commission (1958) and the Rietveld Commission (1970).
[8] Commission on the Concentration of Public Utilities (in Dutch, **Commissie Concentratie Nuts**bedrijven).

reducing the amount of horizontally integrated distributors to about 20–25. The sector was less in favour of concentration and opted for the following guidelines for integration:

1. a minimum of 30,000 customers for integrated firms for gas and electricity distribution;
2. a minimum of 75,000 customers for electricity-only distributors; and
3. a minimum of 30,000–50,000 customers for gas-only distributors.

These figures resulted in a desirable number of 60–70 companies to distribute electricity and gas (Brandsma, 1985, pp. 14–15).

An important milestone in the reform process was the initiation of the Electricity Act in 1989 for the production sector, which also had several consequences for the distribution companies. In several aspects this Act governed the voluntary process of restructuring by the sector itself. First, a vertical disintegration between production and transport on the one hand and distribution of electricity on the other hand was enacted. At the national level, this clarified the division of tasks in the utility sector. Secondly, the act prescribed a minimum scale of electricity production of 2500 MW production capacity. This minimum capacity was perceived to be necessary to achieve economies of scale. Thirdly, the Act allowed production companies to optimise their production processes according to cost criteria rather than to political considerations of the public owners.[9] The act did not attack the public ownership of the production companies.

Between 1985 and 1989, the emerging reform of the utility sector was a manifestation of a reorientation of the public tasks of utilities. The main question that guided the process was how to integrate the public tasks, obligations and responsibilities of the utility sector with the political wish for an effective and efficient economic performance of that sector. The sector had to meet its public responsibilities, such as reliability and security of electricity supply,[10] at the same time improving the economic performance of the sector. This was not an easy job, because the economic and political conditions for production and distribution had changed. Due to environmental considerations, the sector was facing a new challenge: improving its economic performance under more strict environmental regulations. Actually, the environmental impact of electricity production penetrated the sector during the mid-1970s and cumulated as a significant public issue by the end of the 1980s.

---

[9] Out of the 15 original production companies, about seven were organised as provincial or municipal service departments and the others as stock companies, with provinces and municipalities as the only stock-holders.

[10] See, for example, Article 2 of the Electricity Act 1989.

The first oil crisis redirected Dutch energy policy, with its annually increasing energy supply and demand scenarios. Energy conservation and the diversification of fuels became the main policy themes. For the producers of electricity, these adjustments in energy policy had significant consequences. In the short term, the producers were forced to adjust existing production units to make them suitable for coal firing. Regarding the longer-term energy policy, there was a common political wish to increase the share of coal-fired electricity production. Simultaneously, the environmental impact of the production of electricity (especially with coal) became more manifest. The technologies necessary to limit the environmental impact were not available, and the production sector could not wait for these new technologies. They had only one option: increase the share of coal-fired units, wait for new technologies and, for the short term, resist increasing environmental demands.

The environmental issues the Dutch producers were facing concentrated on the emission of SO_2 and NO_x. At first, the sector resisted increasing environmental demands because this would increase the price of electricity too greatly. For several years, the production sector hid behind the unavailability of technology. However, the environment became a main issue on the Dutch political agenda. The sector was facing an increasing front of political support for more protection measures and better technologies. For several years the sector resisted, for example, the introduction of installations for the desulphurisation of stack emissions. To break the resistance of the production sector, the Dutch government supported financially the research and development of new techniques such as burning techniques with low NO_x emissions, coal gasification and the like. In the 1980s, the technology for desulphurisation became available, and the production sector had no arguments to resist its introduction. Also, the Dutch government had supported financially the development and introduction of this technology. Now the environmental impact of coal-fired units is managed by new technologies such as coal gasification and added technologies to clean emissions of SO_2 and NO_x. In combination with quality restrictions on the fuel, for example the amount of sulphur in coal and oil, these technologies limit the environmental impact of the production of electricity.

In between, the Dutch policy on fuel diversification was redirected once again. Consequently, natural gas again became available as a major fuel for the production of electricity. Gas became the dominant fuel for the production of electricity. The electricity sector made an agreement with central government to prefer gas above oil. In this way, central government was able to get more extended revenues from the gas trade in The Netherlands, which eased their budgetary problems.[11]

The re-introduction of natural gas as a fuel for electricity production enabled the sector to meet the more restrictive environmental regulations, at the same time improving the economic performance of the system. The greater degree of cost efficiency of the Dutch electricity sector is partly due to the low price of natural gas.

The reconsideration of the public task of the utility sector concentrated on the reliability and security of supply. Security of supply had to be guaranteed by a system of central planning of production capacity. The Electricity Act prescribed a planning system which was guided by the sector itself, based on a forecast of energy demand over a period of ten years in advance. The sector is legally obliged to actualise these plans every two years. These so-called electricity plans (in Dutch, 'Electriciteitsplan') have to be approved by the Minister of Economic Affairs.[12] This system of centrally approved production planning was meant to protect consumers and to secure a reliable electricity supply. The introduction of this system of central planning provided central government for the first time with an effective instrument for energy policy. This governmental involvement in the utility sector can be perceived of as a form of self-regulation which is controlled by a central authority. The same happened in the case of the protection of small consumers. The Dutch government agreed upon maximum tariffs with the utility sector. The production costs were pooled according to a formula in order to compensate the production costs of 'reasonably efficient' producers.

Between 1985 and 1989, the utility sector went into transformation, securing its public task by central government's supervision. Meanwhile, the sector was allowed to restructure institutionally. During

---

[11] The Dutch government gets about 80% of the revenues of the gas trade from Dutch fields. The other 20% is for the exploiting oil companies (mainly Exxon and Shell). The main gas producer in The Netherlands is the NAM (a joint venture of Shell and Esso) which is, among others, responsible for the production at the big Groningen gas reserve. All producers in The Netherlands (including offshore production at the Dutch part of the North Sea) deliver their production to the Gas Unie (50% controlled by the Dutch state and 50% by Shell and Esso). The Gas Union is responsible for the transport and storage of gas in The Netherlands, for the gas exports to, for example, Belgium, France and Italy and for (future) gas imports, for example, from Norway. The Gas Union has also developed a long-term gas delivery plan, addressing the exploitation of the reserves, importation and exportation issues, and the development of infrastructure, for example, the building of gas storage facilities to meet peak demand in consumption. About 47% of Dutch gas production is exported, and about 5% of Dutch gas consumption is imported.
[12] This right of approval was agreed upon between the sector and the Minister of Economic Affairs in a convenant in 1975. This is another example of the Dutch consultation economy.

these years, the production sector merged into five giant production companies (after 1989 into four), united in SEP, which co-ordinated their production activities to improve the economic performance of production. After the reorganisation, the four producers all had the legal form of private stock companies, allowing only public organisations as stock-holders.[13]

The distributors also merged, a process which turned out to be very successful after a reticent and sluggish start. The process developed almost perfectly according to the scheme provided by the commission installed by the sector (the Brandsma Commission). The big provincial distributors became the most important buyers of the smaller municipal utilities. These provincial distributors, which were legally separated from their production activities by the Electricity Act, achieved good financial positions due the relatively high degree of the accumulation of equity. The provincial stock-holders only required them to pay dividends as high as the rents on the capital market plus a few per cent premium, whereas the old municipal utilities had to pay 75–100% of their annual surpluses to their municipal owners. The big provincial distributors offered very attractive takeover prices to the municipal owners, which many of them gladly accepted. Some municipalities also received shares in the provincial distribution company and/or seat(s) on the board of commissioners. Once the first deals were made, the pattern was imitated, sometimes even under better conditions, reorganising the distribution sector towards increased concentration.

In this intermediate period between 1985 and 1989, a reorganisation process was initiated which for decades had been thought to be impossible. By the end of 1989 the organisation of the sector was roughly according to the agreements made between the sector and central government. However, the process contained an internal dynamic which surpassed the official objectives of the reorganisation process, initiated in 1985 by the sector itself following the objectives of central government to improve the economic performance of the sector.

III.C. 1990 to 1995: Market upheaval

Between 1990 and 1995, the electric utilities developed the attitude of self-assured business firms. The Electricity Act legalised possibilities to trade electricity with auto-producing industrial users. Vertical disintegration freed the distributors from the dominance of the producers. The

---

[13] Stock-holders are provinces, municipalities and distribution companies. Not all distribution companies are stock-holders of electricity producers.

merging and ever-growing distribution companies strengthened their status as important and dominant market parties. Like private business firms, they took the opportunities offered by the Electricity Act to start to explore new markets and to offer new products. However, the new institutional arrangements turned out to be, in some respects, destabilising and threatening to the existing economic order of the sector.

In the first half of the 1990s, the distributors became the dominant actors in the Dutch electricity sector, a dominance which nobody forecast, but which was embedded in the Electricity Act. Before the reorganisation, the big, vertically integrated producers/distributors were the most important economic actors in the electricity market. The mostly small municipal distributors were obliged to buy all their electricity from their regional producer. These producers sometimes also decided the tariffs for consumers which 'independent' municipal distributors were allowed to charge. The reorganisation changed the market position of suppliers and buyers in several respects. Due to the concentration process, distributors had a significant market position because the number of actors was drastically reduced and the purchase volume of each of them grew significantly. The producers depend economically on the distributors because they lack the legal possibility to sell electricity directly to consumers. Only very big industrial consumers are allowed to bypass the distribution companies. Due to the abolition of the regional monopolies of the producers, the distributors could compare prices. Shortly after the introduction of this possibility of 'horizontal shopping', all significant price differences between the producers vanished, and no distributor used this new possibility of free purchase within the Dutch territory.

The possibility of small-scale electricity production which was granted to distributors by the Electricity Act was used by them and resulted in a significant competition with the big producers, united in the SEP. In latter years, the distributors invested heavily in decentral electric power production with small-scale co-generation units, partly as joint ventures with private industrial firms. These activities were stimulated by the Electricity Act of 1989 and achieved societal and political support for environmental reasons. In this way they compete with producers in a significant way. Meanwhile, about 20% of the national electricity production[14] is decentrally generated.[15]

For environmental reasons, the distributors obtained strong political support for their production activities with CHP and renewables. The

---

[14] *Elektriciteitsplan 1993–2002*, Arnhem, pp. 14–18.
[15] Distributors are allowed to install decentral small-scale production units according to the Electricity Act of 1989.

technique of co-generation, generating combined heat or process steam and electric power, is well known and applied by industrial producers in The Netherlands. In the past (until about 1986), only large industrial producers with concurrent heat and power demand and a high load factor used co-generation techniques. Also, in some cities co-generation is used for district or city heating. A critical factor for co-generation was an almost complete use of the electricity generated inside the company because of the low tariffs for surplus electricity produced being delivered to the electricity distributors/producers. The application of co-generation results in substantial energy conservation. Therefore, during the 1980s, the government and the distributors became interested in the extension of co-generation. Also, the tariff for electricity delivered to the distributors has risen from a price based solely on avoided fuel consumption to a price based on avoided fuel consumption plus diminished power demand. A subsidy program for co-generation has been implemented. Presently, distribution companies invest directly in small- and medium-scale (<25 MW) co-generation plants, and also participate in joint ventures with industrial companies building bigger co-generation plants (up to 400 MW).

During the 1980s a governmental stimulation program was developed for renewables, based on substantial investment subsidies. The objective of this program was to expand the installed power of wind turbines and to build up a cost-effective wind-turbine production sector. Until recently, the distributors were reluctant to incorporate wind energy. As a part of their environmental programme, they are now investing in large-scale wind-turbine 'farms'. In the 1980s the Government stated the installation of 1000 MW of wind power as a long-term objective for the year 2000. This objective is reduced to 550 MW in the year 2000. By 1993, about 100 MW had been installed.

Distributors also expand in new markets. Recently, they established a telecommunication company which will compete with Dutch telecom (PTT).

Concluding, it can be stated that after 1990, distributors took several opportunities to expand their position in the electricity market, partly on account of the position of the producers. They penetrated the production market, supported by the environmental considerations of the Government. The institutional framework which was erected by the Electricity Act turned out to be hardly suitable to cope with these unpredicted developments in the market. The central planning system of the producers, for example, was seriously threatened by the production activities of the distributors. The distributors used the financial advantages of decentral CHP production while burdening the central electricity producers with the burden of a growing overcapacity. The reorganisation of the electricity market, formalised in the Electricity Act, made the distributors a dominant economic market party, a position which undermined the institutional framework and resulted in market upheaval in the 1990s.

III.D. 1996: Development towards a liberal market regime?

As mentioned in the introduction to this chapter, the Dutch government very recently produced a policy note about the liberalisation of the electricity market which is intended as a blueprint for the future structure of this sector. Although the note has still not been officially approved by Parliament and is of a quite general nature, it provides interesting insights into possible future developments in the electricity sector.

The government generally proposes free access to the gas and electricity grids, based on non-discriminatory conditions. However, this right to free access will be granted to different groups of customers in different time schedules. Three groups of customers are identified:

1. Captive customers which consume annually less than 50,000 KWh of electricity or 170,000 m^3 of gas. These captive customers will be provided by the regional distributors as natural monopolists. They will get access to the grid as a last customer group, when it seems appropriate.
2. An 'intermediate group' of customers, with electricity consumption between 50,000 and 10 million KWh per year, or between 170,000 m^3 and 10 million m^3 natural gas. This group will be captive for a maximum period of five years. Exemptions for earlier access might be possible in the coming period.
3. Very big industrial consumers with more than 10 million KWh of electricity consumption and more than 10 million m^3 of gas. This group will have immediate access to the grids.

Free-access customers have a completely free choice of energy producers or traders. In this market segment, the price mechanism will be assumed to guide economic activities.

The captive customers and the 'intermediate group' are provided with electricity by the existing distribution companies, which have a legal obligation to serve them. Distribution companies are free in their choice of electricity supplier, and thus might also get involved in direct international trade. Captive customers will be protected in several aspects. In order to guarantee safety of supply, distribution companies have to prove to the Government that they have contracted sufficient capacity to meet demands. Also, the tariffs charged to these consumers have to be approved by the Minister of Economic Affairs. In this market segment, the controlled self-regulation of the present system will basically be continued. However, the Government quite strongly intends to reduce this market segment, and finally grant all consumers free access.

The four existing electricity producers are intended to merge together into one big firm. This concentration process is thought to be necessary to provide Dutch electricity production with a stronger position in the anticipated liberal European market. Competition from the much bigger French and German companies appears to be less threatening if there is one large-scale Dutch producer.

The national electricity producer will be organised as a private-law stock company, with distribution companies and some big municipalities as the only shareholders. Thus, there is some vertical integration by way of the ownership structure. This point of the independence of the electricity producer is at the moment under political discussion. It remains to be seen how independent producers and distributors might act on the liberal market share. It might be that in the future, electricity production is privatised in the sense that private investors are allowed to buy shares in this firm. It is intended to maintain the vertical disintegration between production, transport, distribution, and in future the trade.[16]

The dominant instrument of central-governmental energy policy, the 'Elektriciteitsplan', will be abolished in the new regime. Instead of this, it is planned to submit an Energy Report every four years to Parliament, in which important developments in the electricity and gas markets will be presented and analysed.

As a general impression, it can be stated that there seems to be a genuine interest on the part of the present Dutch government and certain market parties in the development of a liberal electricity market regime. Sections IV and V will elaborate on the interests of different stake-holders in this sector. However, it remains to be seen in the coming years how market parties behave, and whether some of them will act to protect their monopoly positions.

IV. Evaluation of the Reform Results

The results of the reform of the Dutch electricity sector will be evaluated as of the end of 1995. New proposals by the Dutch Government to further reform have not yet been implemented, and thus it is not appro-

---

[16] The gas market will be liberalised roughly along the same lines.

priate to include them in the analysis. The evaluation is based on the following criteria.

1. The stability of the present institutional framework. It will be evaluated whether the industrial structure is satisfying for all important market players, or whether they have incentives for modifications of the present institutional settings.
2. The economic performance of the sector. An important objective for the reorganisation of the electricity sector was to enhance the economic performance in terms of lower tariffs and better economic efficiency.
3. Fulfilment of public tasks. The electricity sector has to take care of certain economic and social goals which are often described as 'public tasks'. In The Netherlands, politically sensitive subjects in this field are the security of supply and protection of the environment.
4. Innovativeness. To improve the economic performance of the electricity sector, innovations are a necessary precondition. Attention is paid to process innovation in the production sector.

IV.A. Institutional instability

The ongoing reorganisation of the electricity sector has not yet resulted in a stable institutional setting. There are two destabilising factors:

1. the growing economic power of the distributors, who in the current system have no equivalent control of affairs in the sector; and
2. the liberalisation of European electricity markets.

As illustrated in Section III, the reorganisation process resulted in a significant shift of economic power to the distributors, which was neither predicted nor expected. This growing market power, manifested in the much bigger scale and scope of the distributors, is not reflected by the existing institutional structure. Producers still determine the pool tariffs, without the possibility of the direct interference of the distributors. Producers still control the import of electricity for public use. Distributors lack the possibility of free import, as big industrial consumers have.[17] There is a growing resentment of distributors towards this dominant position of producers, which is amplified by EU plans to liberalise European electricity markets. Some distributors feel that they

---

[17] Strictly legally argued, all end-consumers *do* have the right of free import and export of electricity, except for distribution companies.

are able to get better prices and conditions if they have the possibility of free purchase in the European Union. The ongoing process of merger is rationalised in anticipation of the coming liberalisation. As large-scale companies, they expect to be better prepared to compete with other European suppliers of electricity. As a part of this liberalisation process, they want to gain more control on price-bargaining and choice of producer.

The producers want to protect their national market and their influence on the pooling mechanism. Producers are afraid that free access of distributors to international markets will cause a loss of market share for them, and result in excess production capacity and consequently rising costs. Producers would be forced to sell their free capacity on an international market. For the producers, the existing system provides an environment with few economic risks and a great degree of stability. Without heavy external pressure they will not give up this comfortable position in exchange for a competing market structure.

To date, the reform process has not resulted in a stable institutional structure. In the pre-1985 market structure, there was a common interest of the main actors in national planning and co-ordination of the activities in investment, grid and production management. The elements of competition, as a consequence of the reform, resulted in competing interests of producers and distributors. There is no longer one homogeneous electricity industry, but different groups of slightly more competing actors. As stated in Section III, the actual system is no longer able to co-ordinate the production activities of the four big producers, the distributors and big industrial consumers in the most efficient way.

IV.B. Incentives for improved economic performance

There is some evidence that the economic performance of Dutch electrical distributors has improved. Between 1982 and 1992 the productivity of labour in electricity production grew at an average rate of 3% per year, as compared with a national average of about 1.6%.[18] Employment in this industry decreased on a rate of 11%, while production grew by 19%. The tariffs for electricity are almost the lowest in Europe, although significant investments in clean technology were made. After the reorganisation, the distribution companies and giant industrial firms invested in small-scale co-generation units. From an environmental point of view, this is a very good development.

It might seem that these developments turned the reform of the Dutch

---

[18] *Source: Annual report 1993*, NV SEP, Arnhem, p. 11.

electricity industry into a success, but this impression is only partly true. The low costs of electricity production are, for a significant part, due to the low prices for natural gas.[19] In the 1970s, after the oil crises and the very high gas prices, the quite undifferentiated production caused very high tariffs. There is a second reason why the rate of tariffs is a biased indicator for the performance of the sector. The Netherlands is a big importer of electricity in the European Union. In 1993, about 15% of the energy delivered by the pool (SEP) was imported. SEP is taking advantage of the low prices of French nuclear electrical power. In this way, the favourable tariffs are to a significant degree due to external developments which cannot be evaluated as a direct result of the Dutch reform. However, without additional research it is not possible to separate the price effects of the growing productivity of the electricity industry from external developments.

The growing number of co-generation units of the distributors, respectively large industrial producers, is exploited at the expense of the four producers for the following reason. These small-scale units are not part of the national system of production planning and co-ordination. By using their own decentral units, the demand for central capacity is consequently declining. As a result, the percentage of central production capacity which is not used productively is increasing, giving some reserve capacity. In 1993, the production of central capacity declined by 2.5% as compared to the previous year.[20] This declining productivity of capital is a source of inefficiency, and consequently rising costs. The producers therefore argue for an integration of these decentral production activities in the system of national planning.

IV.C. Reorientation of public tasks

As pointed out earlier, there has been a remarkable change in the attitude of politicians towards the public tasks of electric utilities. Currently, public utilities are expected to be managed like private sector firms much more than before.

In technical terms, the fulfilment of public tasks by electric utilities meets public expectations. In general, the uniform consumer tariffs are appreciated, although there remain differences due to the characteristics of the region and the efficiency of the local distributor. The supply of electricity is secure. The grid is managed by SEP, as regulated by the

---

[19] As mentioned in Section II.C, about 60% of the total production capacity is gas-fired.
[20] *Annual report SEP*, p. 28.

Electricity Act. Only very big industrial consumers have had difficulties due to the monopoly power of the producers and SEP to effectuate their right of free import. The producers tried to frustrate free import and free purchase within the national borders by charging high tariffs for using the grid. At last these difficulties could be addressed. As a result, big industrial consumers obtained competitive tariffs from Dutch producers, and consequently it was not necessary for them to buy elsewhere.

As a part of the public task, the whole electricity sector assumed its responsibility for environmental protection and energy conservation. The four large producers invested heavily in additional, more efficient technologies. In 1990 the sector signed a general agreement with the central governmental authorities to reduce the emissions of SO_2, NO_x and CO_2. Producers and distributors have to contribute to emission reduction. Producers contribute by investments in technology, efficiency improvements and the like. Distributors contribute by investments in CHP and by initiating energy conservation programmes for private and industrial consumers. In this way, the major actors of the sector take responsibility for improving the environmental impact of the electricity system.

IV.D. Production innovativeness

In the short and medium terms, the following developments may indicate a greater degree of innovativeness of the production sector since the reform:

1. further penetration of integrated steam and gas (STEG) turbines;
2. introduction of large-scale integrated plants of coal gasification and electricity production;
3. further increase of local combined production of heat and power (CHP) with the help of gas engines and gas turbines; and
4. the introduction of natural gas consuming fuel cells, resulting in local production of heat and power and the international grid integration.

Most of these options are based on the long-term availability of natural gas for electricity production; some of them are based on centralised production, and others are dependent on local conditions (i.e. heat demand).

In 1996, a so-called STEG plant with a capacity of 1675 MW and a designed efficiency of about 55% will result in an increase of the average overall efficiency of the system by 2% (to about 44%). This plant will also be the last of the large-scale STEG turbines to be built in the 20th century. The completion of new large-scale STEG turbines is not required before 2005 according to the most recent Dutch Electricity Plan.

The future construction of plants based on large-scale integrated coal gasification and STEG electricity production out of coal gas will probably result in small improvements in efficiency compared to plants using pulverised coal. This will not affect the average production of electricity. The environmental advantages (almost complete desulphurisation and substantial reduction of emissions of NO_x) of the IGC/STEG concept are more important. After the evaluation of the results of a medium-sized demonstration plant, a final decision will be taken about the construction of the first large scale IGC/STEG plants, to be completed in the year 2002. More IGC/STEG plants are not considered until the year 2005.

The market share of power generated by CHP plants for district and city heating is expected to increase substantially as the market share for power generated by industrial CHP plants. Both the producers and a joint venture of distributors and private companies are planned to increase CHP capacity. Producers (SEP) have planned to built about 2100 MW CHP plants during the planning period of the 'Elektriciteitsplan', and distributors and industrial companies intend a capacity increase of 1350 MW by means of CHP plants.

Substantial application of fuel cells will most probably not occur in the present decade, because of the present state of technology in this field. In the next decade, fuel cells may become applicable in the electricity sector. Fuel cells generating power and heat will be used at a local level, and will compete with CHP plants using gas engines or turbines.

International grid integration, growing co-operation and corresponding information techniques may result in a lower requirement of installed electric power without the loss of security of supply and reliability of the system. For instance, the recently agreed co-operation regarding short-term planning of maintenance and production between SEP and the Belgian national electricity company, Electrabel, resulted in a reduction of the maximum required power of 600 MW in The Netherlands.

To conclude, in the short term (until the beginning of the 21st century), technology will result in a substantial increase of the efficiency of electricity generation.

V. Prospective Reflections

There is still much uncertainty as to where the Dutch reform process will finally end. Without going too much into speculation, this last

Section will sum up some considerations which could be helpful in determining future developments. First, an overview will be given of the orientations of different stake-holders in this sector. Second, some future developments with respect to the Dutch electricity market will be elaborated. Based on this information, a short conclusion shall be drawn.

V.A. Orientation of key stake-holders

The outcome of the reform process depends on the position which key stake-holders might take. Key stake-holders include:

1. producers;
2. distributors;
3. regional and central government; and
4. big industrial consumers (described in The Netherlands as 'giants').

The four big producers are confronted with the dilemma that distributors and some big industrial consumers have created alternatives to their supply by means of small-scale auto-production, without having possibilities to expand into other markets. This imbalance of options unilaterally favours the distributors and the big industrial consumers, who can take advantage of cheaper small-scale co-production, leaving the economic consequences of the inefficient use of large-scale production facilities to the producers. Big industrial consumers even have the possibility of free import, which creates another interesting economic option. Before the reorganisation of the sector, the distributors were contractually obliged to purchase electricity from the local producers.[21] Up to 1995, producers were strongly in favour of a stricter system of central planning in which decentral small-scale production units are taken into account. There are also certain movements to reintegrate production and distribution to achieve a more stable system. Producers emphasise the public tasks of electric utilities, stressing that a system of central planning is necessary to guarantee security of supply, low costs, environmental protection and acceptable tariffs.

The distributors can be considered as the winners of the reorganisation. They gained economic influence and bargaining power. However, they would favour even bigger purchase options by allowing them the

---

[21] In the famous Ijsselcentrale case, one Dutch distributor challenged a big producer in a legal case which was even brought to the European Court, on a charge of misuse of a monopoly position. Although this case started in 1985, and both companies no longer exist, it is not expected that a decision will be made before 1996.

right of direct import.[22] Granting distributors this right of free import would undermine the system of central planning in the electricity sector. A more market-oriented system would have to be introduced. This would be in line with the preferences of the big industrial consumers and some distributors who are strongly in favour of a free electricity market. However, there is also a certain resentment in the distribution sector against too much market liberalisation. A free market-oriented system would abolish the comfortable monopoly position of distributors with respect to captive customers. Also, some distributors on their part are showing interest in working closely with their (local) producers. This might be a means of protecting each other's market position. There are some groups of distributors arguing in favour of vertical reintegration with electricity production.

The position of the central government is clearer, after the publication of the Energy Policy Note, 1996. But, as is good tradition in The Netherlands, further elaboration of this note and the way towards a possible liberalisation will follow consultations with the important actors in this sector.

Industrial users favour a more liberal market regime, but do not publicly take a very strong position. Giant industrial users in the present market structure have a comfortable situation which provides them with very low tariffs. Also, they have the possibility to sell their decentrally produced electricity to the public net. It seems that there is some concern about losing these privileges.

V.B. Future political developments

It turns out that the current system is institutionally weak. In the coming discussions about the Energy Policy Note of 1996 and the legally required evaluation of the Electricity Act of 1989, it cannot be ruled out that certain amendments with regard to the right of free import or the vertical integration will be implemented. In this evaluation process, European legislation will be important. The European Commission is urging the Dutch government to abolish the monopoly position of the

---

[22] There are two activities of Dutch distributors to buy directly from foreign producers, which attracted public attention in The Netherlands. There is a contract with Norway to import hydro-power during the day, and to deliver coal- or gas-produced electricity during the night. The contract will be effective by the turn of the century. One distributor is evaluating possibilities to buy electricity from Iceland, which is produced by thermal power. However, this plan is controversial for technical and economical reasons. There are doubts whether it is economical and technical feasible to transport electricity over such large geographical distances without losing too much energy.

producers, and to allow distributors free access to international markets.[23] A new Distribution Act, which would have to legalise the current organisation of this sector, is still to be passed by Parliament. This creates an institutional setting which is open to several different evolutions. Roughly, two scenarios are possible, the first of which is a market-orientated institutional setting, in which distributors and big industrial consumers have free access to national and international markets. In this situation the monopoly position of the distributors could be weakened by allowing a bigger group of industrial consumers—and at the end, possibly even private consumers—market access. Therefore, there is some reluctance in the distribution sector to promote the free-market idea too much. This could undermine the quite powerful position distributors have under the current conditions. In a free market, distribution and trade would have to be vertically disintegrated. Traders, possibly founded by the traditional distribution companies, would have to compete with other marketers. A second possible scenario is to protect electric utilities in The Netherlands as much as possible against competition from other European countries. This could be accomplished by a stronger co-operation between producers and distributors, which would practically mean a vertical reintegration. Although this development is not very likely to occur after the publication of the Third Energy Policy Note of the Government, there are movements in The Netherlands in this direction. By a process of further merger and integration, the number of market parties in the distribution sector could be reduced to 5–10. This would make it much easier to control the Dutch market by explicit or tacit collusion. Although the precise outcomes of the ongoing reforms of the Dutch energy market are difficult to predict, some kind of liberalisation seems inevitable.

V.C. Long-term technological conditions and opportunities

During the next ten years, the potential for fundamental technological change in the Dutch electricity sector is very limited. The absorption of a few GW of CHP power during the 1990s was a substantial threat for the functioning of this sector. In the period 2005–2010 about 5 GW of conventional power (77% gas, 10% nuclear and 13% coal) will be closed, and will have to be replaced according to the present scenarios. So, in the long run, new opportunities exist for the technological and institutional restructuring of the electricity sector.

---

[23] Recently, a legal procedure about this subject was started against The Netherlands in the European Court.

Reasoning from the point of energy efficiency, it seems logical to expand as far as possible the share of CHP power and fuel cells, if available, in the long term. A substantial reduction of energy consumption and CO_2 emissions will be the consequence of such an expansion. Investments in CHP plants may be initiated by the producers on the one hand, and joint ventures of distributors and private firms on the other hand. Three factors may limit a substantial expansion of CHP power and fuel cells: the availability of natural gas in the long run, technical standards for the reliability of the overall system, and the necessity of long-term delivery obligations between electricity producers/distributors and heat-using companies (as is now anticipated by the market parties).

A further expansion of CHP power in the next century will be feasible only if the availability of natural gas is guaranteed below a maximal price level (175–200% of the coal price); otherwise large-scale coal-based options such as electricity generation with IGC/STEG and coal gasification for delivery to local consumers will be more cost-effective. The introduction of fuel cells will not be limited to the same extent as CHP, because of the potential to produce H_2 by coal gasification, which can be used directly in fuel cells instead of reformed methane.

A further expansion of the decentralised production of electricity may limit the technical reliability of the overall electricity system because of the diminishing span of central co-ordination and control. Central co-ordination of production in local plants will limit the possibilities for the co-ordination of production for local heat-consuming industries, or will create a drawback with regard to energy efficiency because of a partial decomposition of the production of power and heat when required by central co-ordination. The instalment of reservoirs for heat on a local scale may solve, on a daily averaged basis, the problem of matching local demand for heat and central demand for electricity. The introduction of load-management options on the national base (such as switch power which can be switched on and off by central co-ordination, for example, for cooling and washing appliances in households) can match, on a daily averaged basis, the supply and demand of electricity. To solve problems of the imbalance of supply and demand of electricity on a weekly or seasonal basis, and to incorporate fully sources of less predictable power (wind and photovoltaic energy), reservoirs for storing electricity are required. These reservoirs can be found on a European scale by the coupling of the Dutch electricity system to hydro-based electricity systems creating mutual benefits, for example, the production peaks of CHP power supplying heat to households during the winter period coincide with shortages of hydropower in that period. Creating local reservoirs and demand-management options, and realising access to foreign hydro-based reservoirs, will result in additional investments.

The ongoing liberalisation of the energy market may directly limit the opportunities to create long-term mutual obligations between electricity producers and local heat-consumers. Indirectly the spirit of the liberalised market may cause an inclination for industrial producers to become footloose and to refrain from long-term obligations which limit the flexibility of production. In this way, investments in CHP may be limited.

From these considerations, two extreme scenarios are derived. The first scenario is based on the supposition of the availability of natural gas at sufficiently low prices. This scenario results in a share of decentralised production by CHP and fuel cells after the year 2010 above the 50% of Dutch electricity production; the remainder will be delivered by highly efficient STEG and IGC/STEG plants. As noted before, this scenario is neutral with regard to the institutional structure.

The second scenario is based on the opposite situation, i.e. relatively high prices for natural gas or a restricted availability of natural gas for the electricity sector. In this scenario, coal will become the most important source for electricity production (above 50% in the year 2010), and the decentralised production of electricity will be reduced instead of expanded. In this scenario, investments in nuclear power again become probable. The opportunities for fuel cells will remain; however, they will not be fuelled by natural gas, but by hydrogen produced by coal gasification or by photovoltaic power. In this scenario, the central producers will regain the lead position in the electricity system.

VI. Conclusion

The reform of the Dutch electricity market, as initiated by the Electricity Act of 1989, has not yet resulted in a stable institutional setting. This chapter has illustrated that there is an ongoing process of economic restructuring and institutional change. The outcome could be twofold:

1. a reinforcement of the existing system of central planning by integration of the decentral production capacity in the national system of investment planning (the 'Elektriciteitsplan'); and
2. the initiation of a market-oriented system of production and trade in electricity.

The second option seems the most likely to occur. At the moment, there are discussions within the sector, and within the government, as well as between both, about the most favourable institutional setting. By spring 1996, the Second Chamber of the Dutch Parliament will have discussed the organisational structure of the electricity sector as a part of the evaluation of the Electricity Act of 1989 and the proposals of the Government contained within the Third Energy Policy Note of 1996. This discussion will most probably give some more indications about future developments of the Dutch electricity market.

Literature

Baake, H. (1988) *Normalisering van bedrijfswinst, winstpunt?* s'Gravenhage.
Brandsma (1985) *Reorganisatie nutsbedrijven*, Arnhem, p. 9.
Simons, D. (1939) *Gemeentebedrijven*, Alphen aan de Rijn, Dissertation, Rijksuniversiteit Leiden.

Part Four

Public Service Orientated and Cartellised Monopolies

Chapter VII
The French Electricity Regime

MARCELO POPPE AND LIONEL CAURET

I. Emergence of the French Electric System

I.A. The early period

The electrification of urban areas has progressed in France since the end of the 19th century thanks to the dynamism of industry managers. They initially built power stations for their own industrial needs, but rapidly realised the advantages of associating the self-production of power with electricity distribution. The first market, captured from the gas supply industry, was that of street lighting under municipal licence. This service gradually paved the way to electricity distribution to households. Urban electricity distribution constituted a way for market expansion and allowed a better profitability of the developed power infrastructure. Initially, these grids had a limited expansion because their main function was to transmit the surplus of industrial power stations, but power producers soon became aware of the real dimensions of this emergent market. Thus, the industrial activity of power generation and distribution, previously considered as a secondary concern, became a fully independent electricity supply industry, draining important financial resources.

Due to the lack of specific legislation, the first electricity developments took place in a fairly anarchic environment. The power plants escaped any public regulation (except a law concerning insalubrity) and electricity distribution operated under a large variety of licence regimes. Appeal to the private sector was then the only legal alternative, by the application of the 1791 law concerning trade and industry freedom. However, since 1900, the French Council of State ('Conseil d'Etat') has authorised municipalities to generate power, but due to the size of the investment necessary to create a complete generation–distribution network, private capital remained dominant.

The next step was a law promulgated in 1906 which regulated distributors' activity and gave the electricity licence authority to municipalities (even if in some cases the State took the place of local authorities). Nevertheless, a sample concession contract was imposed, which had been approved by the Council of State, defining the rights and duties of the utilities. At the end of the concession contract, municipalities became the owners of the facilities. Thus, the electricity supply industry progressively passed from an exclusively private law activity to a regulated and administrated activity, and the concept of public service gradually emerged.

The matter of power production began to be dealt with only in 1919. At that time, a preliminary text ensured that, in order to install hydropower stations, an authorisation (self-production plants under 500 kW) or a licence (plants over 500 kW or all facilities mainly aimed at public service) were necessary. The first legislation controlling thermopower stations was promulgated in 1935.

The electricity supply industry expanded very quickly from the beginning of the century, and in 1914 there were about 20,000 licences in France to private firms, against only 5% of electrical customers supplied by 250 municipal public corporations ('régies'). Specifically, there were 200 generation companies, 100 transmission companies and nearly 1200 distribution companies (or generation–distribution companies), which generally had various concessions (street-lighting, public gas and electricity supply and tramway lines).

However, industry growth was based on urban consumption, which was immediately profitable. The rift which developed between urban and rural communities worried politicians, but only in 1923 was a specific system of grants for electrification of rural communities applied. The process was reinforced in 1933 with, the institution of the 'Fédération Nationale des Collectivités Concédantes et Régies' (FNCCR), a gathering of local authorities and municipal public corporations. The institutionalisation of urban/rural solidarity was achieved in 1938 with the establishment of the 'Fonds d'Amortissement des Charges d'Electrification' (FACE), which was supplied by an electricity tax applied both to urban (mainly) and rural (marginally) customers.

Moreover, the electricity supply industry had structural problems: managers were blamed for having undervalued the hydroelectric power potential of the south-east, and for having neglected the development of the transmission network. As a consequence, excess coal consumption was induced in northern thermopower stations. In 1938, the French government decided to apply a large sectoral plan, but World War II intervened and the plan was withdrawn. However, coal rationing forced the establishment of a national dispatching centre in Paris in 1939, in order to rationalise the installed power capacity. This was the starting point for the interconnection movement which characterised the post-war period.

At the end of the war, 60% of the power was distributed by 34 out of a total of 1150 companies. The five largest groups generated three-quarters of the power, and the ten main ones generated 90% of the power, out of 154 generation companies. Some of these companies together started to build a national grid under the auspices of the 'Electricity Industry Union'.

I.B. The 1946 nationalisation law

At a time when the main objective was to rebuild the French economy after World War II, the French Government decided upon the creation of a national public company, 'Electricité de France' (EDF) in 1946.[1] The main aim assigned to the new company was the reconstruction of the French electric system, and since its foundation the company has bene-fitted from a quasi-monopoly of generation, transmission and distribu-tion of electricity in France. Indeed, the 1946 nationalisation law expropriated the private power companies but maintained the munici-pal power corporations.[2]

As far as generation is concerned, but four types of power stations could be owned by other producers:

1. Some large units, owned by communities or industries before 1946 and mainly used to satisfy their own needs (if they did not dis-turbed the public grid).
2. Industrial self-generation units, integrated in the industrial process (steam, gas and heating systems).
3. Urban-waste power plants (since 1949) and district heating plants (since 1980) owned by local authorities.
4. Small independent units with an annual output lower than 12 Gwh and an installed capacity lower than 8 MW.

The nationalisation law also defined the purchase conditions for the

---

[1] From the outset, the government recognised FNCCR as a special partner of the electricity industry. It provided three out a total of 12 members in the Administrative Council of the State-owned company. In the same way, local authorities had 12 members in the Gas and Electricity Council of the French Ministry of Industry.

[2] Régies, SICAE (*'Sociétés d'Intérêt Collectif Agricole'*—legal status created in 1920 and com-pleted in 1961, in order to facilitate the activity of rural electrification co-operatives) and SEM (companies combining public and private ownership).

excess power for self-producers and independent power generation.[3] Until 1995, the public utility was obliged to buy this production. Last year this obligation was suspended for three years, except for power generation based on renewables and combined heat and power (CHP). The purchase price is based on the high-voltage tariff, minus an estimated utility-distribution cost. This means that the purchase price is calculated, case by case, on the basis of the costs avoided by the utility. It is up to potential investors to analyse their own interests before deciding to go ahead.

The transmission network was entirely nationalised by the 1946 law. Even high-voltage lines owned by the National Railway Company (SNCF) and the National Coal Company ('Charbonnages de France') were transferred to EDF. The public utility was allowed to transmit electricity from self-producing power plants to another facility within the same company, the head office of the company, or a subsidiary company, but not to more than three points or to other customers.

As far as distribution was concerned, the nationalisation law preserved the distribution co-operatives and other municipal utilities which were already in existence.

I.C. *Public service principles and tariffs*

In spite of the efforts since 1935 to homogenise tariffs, a large pricing discrepancy between the 1200 existing utilities still remained at the end of World War II. So, in 1946, the new State-owned monopoly faced a double economic challenge: on the one hand, the need to provide economic tools to optimise investment choices and, on the other hand, the obligation to build a new pricing structure consistent with the main principles assigned to the public company, i.e. efficiency and equity. A particular pricing system based on marginal costs was set up in the 1950s and 1960s. It is still being implemented and improved upon.

The economic efficiency of the French utility is not just an imposed balance between revenue and expenses, but also integrates the willingness of the collective optimum. It means that utility and State interests are combined via a complicated procedure in order to choose together the best generation/transmission planning at the least cost, develop adapted electricity uses, manage demand according to supply costs, etc. Theoretically, this efficiency revealed a straight and unavoidable link between individual choices and collective costs. Tariffs have been con-

---

[3] The agreement between the State-owned company and independent producers must last for a minimum of five years.

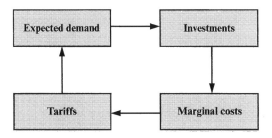

Fig. VII.1. Theoretical tariff assessment.

sidered as this link in order to inform the customer of his or her own real consumption costs for both the electric system and society. Which type of costs should be selected? As a result of the work of neo-classical economists and Maurice Allais's research on 'social output', French economists in the 1950s chose time-based tariffs reflecting future long-term marginal development costs,[4] giving customers good information which enabled them to take the decisions which finally were the best for society as suggested by the well-known Boîteux phrase: 'Tariffs are made to tell costs like clocks are made to tell time' (see Puiseux and Boîteux, 1970).

French tariffs are also based on the price-equalisation principle, which is a social tool for equity and for homogeneous national development. It means that pricing is geographically uniform over the country for the same use at the same time. For a similar electricity use, customers in Paris, in the Alps or in Guadeloupe pay the same price. However, this social choice hides the cost disparities of distribution or generation depending on location, and induces financial transfers from one area to another.[5]

Even if the French electricity-pricing system is considered to be highly efficient, its theoretical and practical limitations are well known, and include the budgetary constraint, which induces a second-best optimum (the Ramsey-Boîteux theorem[6]), the feasibility of tariffs,[7] the limited rationality of the agents (the low share of electricity in the overall expenses of most companies and householders sometimes makes the

---

[4] Supply costs for an additional energy unit. Tariffs based on instantaneous marginal costs would not be stable enough.

[5] Crossing subsidies from urban to rural areas or from continental France to French overseas territories.

[6] This theory balances future-based incomes and past-based expenses.

[7] Theoretical tariffs must be simplified so as to be applicable and understandable by the customers.

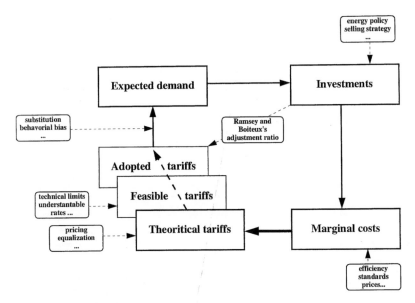

Fig. VII.2. Real tariff definition.

electricity bill trifling),[8] political bias (electricity supply used as a social, political or economic tool), the difficulty of obtaining the best-suited generation mix as defined by the theory,[9] and some remaining market failures. Thus, theoretical tariffs have to be simplified and adjusted.

Furthermore, even under social constraints, tariffs can sometimes remain a strategic tool to develop sales, mainly for competitive uses of electricity such as heat production to industry, tertiary and household customers (process heating, space heating, water heating, cooking).

II. Overview of the Present French Electric System

The present French electric system results from the 1946 nationalisation law, which remains in effect. Only a few changes have been implemented.

---

[8] We can add that the higher the supply quality is, the more invisible it becomes, and the more it is taken for granted (this is a paradox for energy efficiency).

[9] The best-suited generation mix equals short-run and long-run marginal costs.

II.A. State as owner and regulator, and the utility

The Ministry of Industry, through the General Directorate of Energy and Raw Materials (DGEMP) and more specifically the Directorate of Gas, Electricity and Coal (DIGEC), is directly in charge of the French power sector, as a regulatory authority. Other ministries are also involved in legislation and regulation, such as the Ministry of the Economy (strongly involved in tariffs), the Ministry of the Treasury (debt) and Ministry of the Budget (dividends to the State). Since 1982, the relationship between the State (as owner and regulator) and the State-owned utility has been defined by 'contract plans'. A four-year negotiated agreement has been signed between the firm and the Government which defines medium-term objectives. This regime gives the utility real autonomous management, in exchange for negotiated performance targets. The main goals assigned to the firm for the period 1993–1996 are:

1. economic and financial goals: debt reduction of 40 billion French francs (FF) over four years, dividends to the State at a rate of 30% of benefits, and tariff decrease by 1.25% per year in real terms;
2. technical goals: improvement of power quality, increasing the quality of services, and implementation of rational electricity uses;
3. landscape and environment: FF3 billion for burying power cables and FF3 billion more for emission reductions;
4. main company strategies: investment schedule and national and international sale policy; and
5. social policies: employment, wages, etc.

Nevertheless, the main actor in the electric sector is EDF, the public utility created by the 1946 nationalisation law. Indeed, the French electricity generation, transmission and distribution system is characterised by a highly integrated ownership within the public monopoly's hands: 91% of the installed power capacity, 94% of the electricity production; 95% of the high-voltage (HV) transmission grid, and 95% of the electricity distribution.

II.B. Characteristics of the generation and transmission functions

With 54 nuclear power plants operating in 1995, and four more to be constructed before 1999 (representing a further 5.8 GW), France is one of the major nuclear users. Furthermore, its electricity generation mix is very unusual, with 75% coming from nuclear power plants while hydro-power provides 18%. The part of total power production generated by producers other than the public monopoly decreased from 20% in 1970 and 15% in 1980 to only 6% in 1994. The public utility itself plans the day-to-day development of its power plants and transmission network, but the more important decisions are defined in the contract plan. It can be added that since the 1980s, France has excess installed capacity.

The French HV transmission grid (over 63 kV) extends over close to 100,000 km. It is built on a ring-way and is linked with all neighbouring countries and with Great Britain. The public monopoly is entirely in charge of the scheduling and the dispatching of the whole electricity transit.

The generation and transmission function is responsible for 70%

Table VII.1. Thermo-power plants (1994)

	Installed capacity		Net production	
	MW	%	GWh	%
EDF	83,120	90.7	354,725	94.9
Nuclear	64,250	70.1	341,615	91.4
Other thermal	18,870	20.6	13,110	3.5
Others	8480	9.3	19,145	5.1
Nuclear producers*	1490	—	2	—
National Coal Co.	2855	—	8325	—
Industry	4000	—	10,450	—
Municipalities	125	—	368	—
Total	91,600	100	373,870	100

* Belgian, German, Italian and Swiss companies which own shares in some French nuclear power plants.

Table VII.2. Hydro-power plants (1994)

	Installed capacity		Net production	
	MW	%	GWh	%
EDF (including CNR*)	23,030 (3,000)	92.2	72,966	90.5
Others	1950	7.8	7640	9.5
Municipalities	85	—	412	—
National Railway Co.	630	—	1715	—
Industry	195	—	1031	—
Others	1040	—	4482	—
Total	24,980	100	80,606	100

* '*Compagnie Nationale du Rhône*', partially shared by EDF, which manages dams on the Rhône river.

(about FF125 billion) of the total public utility turnover. It must be pointed out that these two functions are both included in the same General Directorate of the utility.[10] With this structure, France is very far from the European Commission (EC) position, which aims at recognising a specific role to the transmission network, which is independent of generation. But, under pressure from the EC, the 'unbundling' of these two functions is expected to occur in France.

II.C. Characteristics of the distribution service

French municipalities are the franchising authorities for electricity distribution.[11] There are three types of power distribution regimes:

1. those entirely controlled by the municipal power corporations (Régie, SICAE or SEM);
2. investment decisions (building and reinforcement) directly controlled by local authorities and management (operation and renewal) delegated to the State-owned utility; and
3. those wholly governed by the public utility under a concession regime.

The last type is usually adopted. For historical reasons, a few towns (such as Belfort, Grenoble, Metz and Strasbourg) and some rural communities still take upon themselves the electricity distribution service. The intermediate procedure is commonly used in rural areas. Nowadays, about 200 suppliers other than the State-owned company are still involved in the distribution of electricity to more than 3.3 million customers (40% in rural areas), representing 5% of total distribution. Inside the company, the 104 public utility distribution centres are in charge of distribution services. Their purchases represent three-quarters of the sales at transmission level, while large industries are responsible for 15%.

In rural areas (concerning 80% of French municipalities), where the FACE nowadays mobilises about FF2.5 billion per year within a total annual investment budget of FF4 billion (to build and reinforce the rural distribution network), a sophisticated institutional organisation has been created to develop electricity supply. As mentioned earlier, the FNCCR is a powerful partner and adviser which helps rural communities and local authorities to manage public services and negotiate with other agents.

---

[10] *'Direction Electrique pour la Production et le Transport'* (DEPT).
[11] Except in some particular cases, where State or regional administration take the place of municipalities as a result of local history.

France has close to 600,000 km of MV (under 57 kV) and about 660,000 km of LV (under 380 V) distribution lines.

II.D. Some characteristics of the electricity consumption pattern

The breakdown of electricity consumption, by sectors and suppliers, is presented in Table VII.3. Following several decades of continuous growth, the average annual increase of national consumption has dropped from 6.5% between 1960 and 1980 to 3.5% between 1980 and 1990, and to only 3.0% in the 1990s.

Table VII.3. Electricity consumption, 1994 (TWh)

	Total	EDF	Others*
Low voltage	140	132	8
Household & agriculture	108	—	—
Professional	32	—	—
High voltage	220	200	20
Industrial	152	—	—
Other	68	—	—
Total	360	332	28

* Including self generation (~ 10 TWh).

Following the classical overall expansion of electricity use, but mainly concentrated in the industrial sector, household consumption has more recently generated a large part of the growth of total power consumption (especially thermal uses like house heating, water heaters, etc.).[12] Today, it represents close to 30% of total consumption. However, this market is now saturated and its potential has been exhausted. Ten years ago 70% of new dwellings were full electricity equipped; nowadays, this has dropped to 50%. Furthermore, the development of electrical space heating has been widespread. Today, 25% of households are equipped with electrical space-heating systems (corresponding to 10% of national electricity consumption). Moreover, its real economic benefit has sometimes been criticised.

Customers' behaviour is oriented by tariffs. An electricity bill is composed of a fixed amount depending on the subscribed power, and a variable term related to energy consumption. Different tariffs are applied to customers according to their power subscription level:

1.   Blue tariff for low-use customers with a subscribed power below

---

[12] Besides some industrial thermal and process uses and, nowadays, air conditioning.

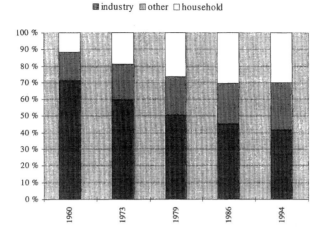

Fig. VII.3. Structure of electricity consumption (%).

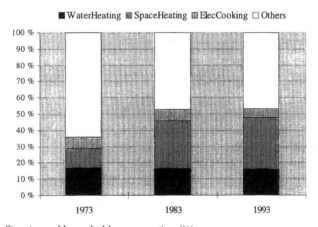

Fig. VII.4. Structure of household consumption (%).

36 kVA: today, there are approximately 4 million professional customers and 24 million household customers.

2. Yellow tariff for customers between 36 and 250 kVA: this applies to 300,000 medium-voltage customers (industries, tertiary, municipalities, etc.).
3. Green tariff for large customers over 250 kVA: there are almost 600 high-voltage industrial customers.

The utility manages the demand through tariff options, like night-and-day tariffs which contribute to reduce consumption at the peak load

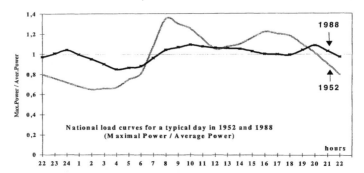

Fig. VII.5. National daily load curves.

and shift the load from the peak hours to off-peak periods. Seasonal tariff options, such as the old EJP option and the new TEMPO option[13] have the same purpose. Tariffs for large customers (i.e. high and medium voltage) have been very efficient in flattening the national load curve between the 1950s and the 1990s. The impact of this tariff policy is illustrated by Fig. VII.5. Nevertheless, besides this effect, the sensitivity of power consumption to temperature has increased significantly in France.[14] Nowadays, more and more real-time tariffs are slowly being extended to domestic customers too, simultaneously with the evolution of metering technologies which can reflect more closely the variations in real-time costs.

II.E. Electricity exports

France exports the equivalent of about 18% of its national consumption and will probably be the only European country with a substantial positive net electricity balance at the end of the century.

II.F. The financial situation of the State-owned company

The 1970s and the first half of the 1980s were characterised by a high level of investment in the nuclear sector, just as the 1950s were character-

---

[13] EJP ('effacement jours de pointe'): for large customers, in exchange for cheaper off-peak kWh, some appliances are switched off during peak days when requested by the utility. TEMPO: for LV customers, it provides 22 'red' days with very expensive kWh and 43 'white' days with medium price/kWh in winter time, and 300 'blue' days with cheap kWh, with night-and-day differentiated tariffs.
[14] Winter peak growth is faster than consumption.

Table VII.4. Net electricity exports (TWh)

	1992	1993	1994
Andorra	0.08	0.08	0.13
Belgium	4.87	2.88	3.40
Germany	2.00	13.37	14.97
Italy	13.00	17.27	17.10
Luxemburg	0.05	0.05	0.05
Spain	0.80	1.59	2.85
Switzerland	13.00	9.19	7.50
United Kingdom	17.00	17.00	17.17
Total	**50.80**	**61.43**	**63.17**

ised by the last large dams. The reduction of the growth in demand, together with the arrival of new nuclear power stations, brought excess installed capacity and high indebtedness. Whatever its financial situation, especially its indebtedness, the State-owned company has always been able to find resources on financial markets.[15] In recent years, the public utility has started reducing its debt, in the context of very low investment levels in generation equipment and slowly decreasing electricity prices.

With about 117,000 employees, and a turnover of more than FF180

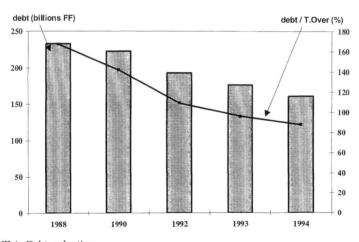

Fig. VII.6. Debt reduction.

[15] With the agreement and the insurance of the Government.

billion,[16] the State-owned company is characterised by a very concentrated and powerful structure. It is also the largest electricity company in the world. Table VII.5 shows the public utility financial results in 1994, and the investment split between generation, transmission and distribution equipment. This is far from the traditional breakdown in the electric sector: 40–60% for generation and 20–30% for transmission and distribution. Today, distribution absorbs close to 45% of sectoral investments in France.

Table VII.5. Financial results in Billion FF

	1994	1995
Turnover	183.3	188.6
Net result	3.2	2.8
Investments	35.1	35.8
Generation	9.7	–
Nuclear	7.9	–
Transmission	5.9	–
Distribution	15.4	–
Other	4.1	
Debt/turnover (%)	87.7	77.1
Investment/turnover (%)	19.1	18.8
Financial expenses/turnover (%)	8.8	6.5

II.G. Other main agents in the electric sector

We indicate below some other powerful institutional or industrial agents linked to the French electricity supply industry in various fields.

The 'Commissariat à l'Energie Atomique' (CEA), set up in 1949 to take charge of nuclear research (both military and civil), has been working in partnership with the public utility since the early 1950s. It was under the double leadership of EDF and CEA that French engineers succeeded in 'frenchizing' an American reactor design, and thus became technologically independent. Due to the strong choice for nuclear power generation in the 1970s, the nuclear sector joined the restricted and powerful industrial group of electrotechnical suppliers which enjoy a strong relationship with the public utility. For example, in 1992, five industrial companies realised FF78 billion with EDF.

---

[16] Nevertheless, in 1994, for the first time, its turnover decreased compared to the preceding year (−0.1%).

Table VII.6. Main public utility's electrotechnical suppliers

	Total turnover (billion FF)	Turnover with EDF (billion FF)	Percentage of turnover with EDF
GEC Alsthom Energy (turbines, etc.)	214	30.0	14
Cogema (nuclear waste treatment)	45	22.6	50
Framatome (nuclear power plants)	28	12.7	45
Merlin Gerin (dispatching, etc.)	44	8.8	20
Pirelli Cables (cables, etc.)	17	4.2	25

The 'Agence de l'Environnement et de la Maîtrise de l'Energie' (ADEME) was set up in 1974 in order to develop energy efficiency approaches and renewables. Since the decrease of petroleum prices in the 1980s, the agency has received restricted means to implement this policy. However, in the 1990s, it has become more and more active in the environmental field, and an agreement was recently set up with EDF to implement demand-side management and renewables.

III. Societal Conventions Sustaining the French Electric System

From abroad, the French electric organisation may seem stationary, opposed to reforms, and sometimes even arrogant. However, in order to understand the French regime, it is necessary to go beyond a technical and economic analysis of the electric system: one must consider the societal conventions which determined the history of the French electricity supply industry. In France, electricity is not a simple good, but also an economic tool, a social symbol, a national pride. It benefits by a quasi-consensus between the State, the company and trade-unions, complemented by a general agreement of the public, as customers and citizens.

III.A. State ownership, public monopoly and public services

The French State has a long tradition of direct involvement in the economy. There is an historical trust in this role. French culture favours the primacy of collective rules over individual achievements. The State is the only embodiment of national interest, and implements the vision of the nation against private interests and their social and political manifestations. Public monopoly is a traditional way to provide public services and the decision-making process is traditionally centralised. As a consequence, in the power sector, the French approach is usually monopolised, monotechnological (i.e. dams, then nuclear plants), and naturally more inclined to implement cost-effective large power facilities within a long-term view than to manage small power plants in order to promote competition. Organised and controlled markets are not discredited, but they correspond to the widespread distrust and dislike of uncontrolled competition, especially foreign competition.

Public utility services are globally considered to be efficient and maintain their positive image, reinforced by people willingness to defend social transfers such as price equalisation. In this sense, the French electric system seems to be the best example of this national satisfaction. The State-owned company is not perceived as a profitable enterprise, but more as a public-service oriented organisation. Its image takes advantage of its historical involvement in the reconstruction of the country after World War II, when it had a major role in the development of the French economy and social stability. Later, the public utility managed the massive investment in the nuclear plant-building program for a relatively short time and with relatively cheap reactors, without any major mistakes. It is the only nuclear power program which is still proceeding, largely unchallenged by the public (except for a similar relative success in Japan). There is today a general pride in French electricity technology, emphasised by no major accident, leading to a wide public acceptance of the nuclear program. However, some expectations about information and 'nimby'[17] behaviour about nuclear waste management can be identified.

Electricity is seen as a public service, which means that all citizens should have access to it in the same conditions (uniform tariffs). Privatisation and a break-up of the State-owned company would mean the end of this nationwide uniformity. This partly explains why those who ask for privatisation of the public utility in France are so few. Nevertheless, this current situation must be tempered. In fact, a reduction of the various monopoly powers in this field cannot be ruled out today.

III.B. Security of supply and energy planning

France has long been obsessed with political and economic independence. In the energy sector, this preference appeared in the past in efforts to build strong national oil companies which could break the

---

[17] 'Not in my back yard'.

stranglehold of American and British oil multinationals on the world oil market. Nevertheless, the oil crisis in the 1970s demonstrated to what a dangerous degree France had become dependent on foreign energy resources, even for electricity generation. This very quickly led to the decision to turn to a nuclear program. The power sector and the nuclear industry have been charged with achieving the national target of decreasing energy dependence. They managed this technological transition quite well, and their capacity to reply rapidly was greatly appreci-

■ coal ▧ oil ▨ gas ■ other □ prim.elec.

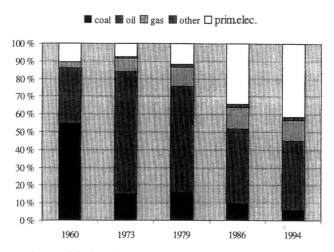

Fig. VII.7. Evolution of the French energy mix.

■ coal ▧ oil ▨ gas ▧ other ▧ nuclear □ hydro

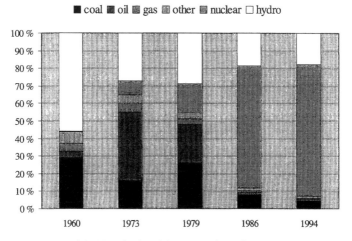

Fig. VII.8. Evolution of the French electricity-generation mix.

ated. The energy independence ratio rose from 22% in 1973 to 51% in 1995, although it is still below the 55% of the European Union as a whole. The contribution of nuclear electricity to this change is obvious when we compare its share in the national energy balance: 5% in 1979 and 35% in 1995. However, one must point out the resulting large power over-capacity.

This strategy was developed in the 1970s and the 1980s with the agreement of the French authorities, who accepted a large period of high indebtedness of the public utility and a very large pay-back period for investments. So, in a context of weakness of environmental sensitivity and absence of a real national debate about nuclear energy,[18] the influence of politicians and technicians was very strong.

The engineer's view must be taken into account to explain the French approach. It is deeply rooted in the elitist system of 'State Corps'.[19] The highest positions in the Ministries and State-owned companies (such as EDF) or in nuclear industry are held by people who graduate from elite engineering schools which preach the public interest ideal. Their ability and efficiency as decision-makers are not in question. The negotiations about energy planning and technical and economic choices are strongly founded on well-developed microeconomics instruments ('calcul économique'). The sophisticated French electricity pricing system was created by these engineer-economists.

III.C. Employment and social gains

The State-owned company remains a symbol of the social gains of the 1950s and 1960s. It has been innovative for more than two decades, by implementing new models of employment and wage policies, welfare benefits and working conditions, retirement plans, etc. The company was an example not only for other State-owned companies, but sometimes also for the private sector. Of course, its 117,000 employees now defend their acquired rights. Trade union organisations, too, cling to the concept of public service, and are obviously defending one of its last real strongholds within the working class and the economy. Employees and trade-unions are active players in electricity debates, and decision-makers have to take their opinions into account.

---

[18] The nuclear program was launched in 1974 with the agreement of the two chambers of Parliament. French citizens and customers were not invited to give their opinion directly, and are usually relatively poorly informed.

[19] Only the ten best engineers from the *Ecole Polytechnique*, then the *Ecole des Mines* or the *Ecole des Ponts* are invited to join the 'State Corps' each year.

III.D. National consensus

The powerful national consensus, as far as the French electricity regime is concerned, results from an interplay over the previous five decades between the three main institutional players in the power field: the State, the public utility and trade unions. French societal conventions explain why a strong movement for economic liberalisation by non-State agents cannot be foreseen in the short term. As a matter of fact, there is no real claim for privatisation or liberalisation of the power sector in France. It provides a high degree of energy independence and contributes to keep main national balances such as the export/import ratio in equilibrium. The public utility has been able to reduce its dangerously high level of indebtedness in the 1990s. It has been profitable since 1990, although its tariffs are decreasing in real value. Furthermore, in the short term, the State-owned company does not need significant financial resources for the expansion of generation, transmission or distribution. With no financial 'throttling', one of the main motivations for sectoral change does not exist in the French case. No major economic or non-economic reason would apparently justify upsetting the existing situation.

A large consensus exists in France to preserve the public-service orientation of the power sector, but some liberalisation could be implemented in the near future in some specific areas, such as power generation and distribution, particularly related to combined heat and power and renewables. New debates have also been launched in recent years on the very notion of public service, the improvement of regulation and energy efficiency, environmental behaviour, and the internal and external role of the State-owned company.

IV. Seeds of Transition in the French Electric System

Even with its recognised efficiency the French electric system is changing, as a respose to present debates and challenges. From abroad, the liberal wave, the restructuring of foreign systems and global debates such as climate change, have not left the French organisation unaffected (involving mimetism, adaptation or defence). In France, the end of the large investment program, the resulting over-capacity, the saturation of traditional markets, the possible arrival of competitors and the emerging environmental issues (landscapes, greenhouse-gas emissions, nuclear waste), have obliged the electricity industry to react for its own development. Moreover, to face recent technological and managerial innovations such as CHP, renewable, demand-side management (DSM), least-cost planning (LCP), integrated resource planning (IRP), independent power producer (IPP), real-time tariffs, spot market, etc.

IV.A. National energy debates: The reports Mandil and Souviron

Since 1993, a real national energy debate has been launched under the initiative of the Ministry of Industry. The first step was the publication of a report, 'Reform of the electric and gas systems', (Mandil's report). This report was clearly an intermediate answer to the European Commission, proposing several marginal reforms while defending the French regime. Its conclusions were not opposed to a breakup of the generation monopoly, arguing that the State-owned company would be efficient and competitive enough to face potential competitors, but that this competition should be controlled to protect the system and its basic principles through a single grid manager.[20] This change would involve new purchase tariffs which remain to be discussed. Since 1946, the public utility has been obliged to buy excess self-generated power from small units (under 8 MW). This obligation has, since 1995, been temporarily abolished (for three years at least), except for independent power production based on renewable energy or for self-generation based on combined heating and power. The public utility can also buy, under certain conditions, classical independent generation during peak-load periods. One of the most important parts of the report concerns the lack of independent expertise regarding electricity costs.

The second step was the first 'National Energy and Environment Debate', which was a year-long regional (then national) meeting regrouping ministries' staff, utilities' personnel, local authorities, experts and citizens. This debate was recorded in the Souviron report. As far as electricity is concerned, the report emphasises the right of each citizen to have access to information, which is not the case today. There has been a lack of democratic debate in the past about the national energy policy, a lack of transparency on security and the real costs of nuclear waste management, on the future dismantling of nuclear plants, and on the real costs of electric heating for housing (which has been particularly criticised). Like the preceding report, this one argues for the development of an effective independent expertise to improve this situation. This evolution would allow the new expectations of people to be included in the local decision process, because today energy is perceived

---

[20] In the short term, it would organise dispatch and collect the cheapest supply from EDF or from independent suppliers after bidding or auction. In the long term, it would control plant programming.

as more than an economic issue. The environmental theme is unavoidable. A few changes have already taken place, such as the audition process for a new HV line project. The 1992 agreement between the government and the public utility to bury LV lines is judged to be insufficiently ambitious. Furthermore, renewable energies should be developed. But French research is mainly focused on nuclear and oil technologies, and renewables are not sufficiently considered in comparison to the case in foreign countries. In conclusion, the report shows that no energy and environment policy can be fixed once and for all.

IV.B. Local authorities: New real partners

Because each local authority signed a distribution concession contract with the electricity distributor, they should theoretically have some power to influence developments. But this was not the case: the State-owned company was in fact the obliged distributor, except in some rural areas; and the previous obligatory contract (agreed by the State Council) was set up in 1928 and was still in effect after the 1946 law, until the 1980s.

Some current institutional events have changed the deal and made the contracts appear out of date. The 1982 decentralisation law gave new responsibilities to local authorities. Nowadays, they are more powerful and accountable for orientating local choices than before. On the other hand, the protected position of the public monopoly began to be contested at the European level. In this context of new responsibilities and pressures, a new concession contract was debated in the 1980s. In 1992, a model contract was adopted, negotiated by local authorities, the Ministry of Industry, the FNCCR and EDF. Franchising authorities added new conditions to the model contract, which is closer to their own interests. It now requires improvements in the quality of service and environmental goals (especially the burying of cables), emphasises energy efficiency, states new utility obligations (particularly concerning yearly reports) and reaffirms the control role of local authorities. In May 1995, 23,000 local authorities signed a new concession contract with EDF, which is applicable for the three next decades.

The new concession contract reveals the opportunity for municipalities to reactivate their responsibilities. Besides, both the Mandil and Souviron reports argue for more and more regrouping of small allowance authorities to make them more influential. For the State-owned company, it was also a good way to appear open-minded and closer to the expectations of municipalities, which are also its customers, and a good way to reinforce its position in the local public service market.

Another local debate is emerging. The implementation of new HV transmission lines is being questioned more and more, and faces the reluctance of local people (translated by the voices of local politicians). Previous debates concerned only the definition of line tracing. Today, the State-owned utility must justify the function and the necessity of grid extension. More and more, independent studies, ordered by local authorities, refute development projects for transmission lines. Independent experts are often consulted and propose alternatives, such as demand-side management, to avoid a new HV line. For example, in February 1996, the decision to construct a transmission line between France and Spain was cancelled because of environmental issues.

IV.C. European Union constraints

European Commission policy-making is presently a complex process and lacks transparency. Currently, several debates are in progress concerning the field of electricity. For instance, with regard to greenhouse-gas emissions, the French position is favourable to taxation based on the energy/CO_2 ratio. The nuclear share in power generation in France means low impact from such a taxation on French electricity prices, and will probably favour the competitiveness of the French electricity industry over producers from other countries. Another debate concerns third-party access (TPA) and competitive markets.

The French State-owned company import/export monopoly was regarded as contrary to the Maastricht Treaty rules. The French reply is usually based on the necessity of maintaining basic national principles such as price equalisation and strategic supply continuity obligation, which are claimed not to be compatible with short-term oriented open markets. The TPA debate was launched nine years ago, but was at first totally rejected in France. In 1994, the Mandil report suggested an alternative called the 'single buyer', applicable to large industrial customers: this was an agreement on competition between electricity suppliers for these large customers, but with the public utility maintaining the central control role of buyer for foreign power (administrative allowance). The French position could reconcile the public utility's need to export nuclear base-load power and its need to import cheap peak-load electricity from abroad, without losing monopoly control over the domestic market.

The recent directive adopted in June 1996 by the European Union Energy Ministers, progressively opening (over ten years) the electricity market for the largest customers, constitutes a new framework for electric systems in Europe. In France, it means initially 400 customers over 40 GWh, then 3000 customers over 9 GWh who will be allowed to choose their supplier (32% of the French market). Both the public utility and the State are officially confident, claiming a cheap and competitive kWh. The future will reveal the extent of the change being introduced today.

IV.D. Potential competition in the domestic electricity market

The main players targeted by the changes debated in the Mandil report are the private franchised water companies. Up to now, EDF has been the main operator in the electricity field. But a few private companies,[21] such as the powerful water distribution companies Compagnie Générale des Eaux (CGE) and Société Lyonnaise des Eaux, have started to challenge its monopoly position, occupying all the free spaces allowed by the law, mainly by implementing a lot of small power generation units (below 8 MW). These companies are traditionally involved in the management of local public services. As a matter of fact, if local authorities usually control public services, such as water or electricity distribution, they frequently delegate management to public service companies (CGE or Lyonnaise for water and EDF for electricity) by signing long-term contractual agreements. From water distribution, these two companies have progressively enlarged their activities to waste management, public works, telecommunications (cable TV, telephone) and other public services, in France and abroad, in the stream of the deregulation and privatisation of public services all over the world. They are leading the world market in water distribution, the European market in waste management and the French market in energy services. In 1995, CGE reached a turnover of FF165 billion and Lyonnaise a turnover of FF98 billion.

Energy activity represents FF33 billion for CGE and FF10 billion for Lyonnaise, even if their electricity share is still very low. Their energy strategy includes energy services (management and maintenance) for municipal equipment, public and private buildings, district heating, etc. Both companies are developing power generation activities, i.e. combined heat and power, waste-to-energy, natural gas, and small hydroelectric facilities. The strategy is based on the management of a wide range of technologies, which are generally not developed inside the companies, but rather under license. This diversified set of techniques makes them able to fit local demand. It also contrasts with the electric

---

[21] Other public and private suppliers exist (see Sections II.B and II.C), but their role remains strategically limited when compared to that of the two main water companies.

system as a whole, which is strongly based on large nuclear power plants. The bases of their electricity business are different. Lyonnaise was largely involved in electricity production and distribution before 1946. Up to now, it has operated local power generation, and distributed electricity and gas in a few towns (Bordeaux, Grenoble, Monaco and Strasbourg). On the other hand, without any history in the electricity field, CGE has developed its experience abroad, as in New York State (USA), where it manages 1800 MW of power plants and earned more than FF3.6 billion in 1995. Today, it also owns power generation units in France and other European countries. Both companies are gaining experience in order to be ready for a possible deregulation of the French electricity market.

IV.E. Diversification strategy of the public utility

The diversification of the State-owned company is now well under way. The turnover of diversified activities in 1993 was FF4 billion, still modest (2%) compared to the firm's total turnover. Since 1991, the public utility has started to extend its knowledge and ability in new fields such as engineering, cartography, TV cabling and waste management. This is a way to remotivate its employees, who are facing slowing electricity demand, and to pose new challenges. This strategy was questioned under the argument that the company could take advantage of its monopoly position to penetrate other markets and develop unfair competition against smaller private companies. Diversification was thus limited, but EDF was allowed to pursue its activities in some domains such as public lighting, water distribution and, mainly, urban waste management and TV cabling. Via its subsidiary TIRU, EDF is now the third operator in the French urban-waste management market, behind the two water companies. TIRU is specialised in the construction and exploitation of urban waste incinerators. The company operates plants in Paris and its nearer suburbs, and one of the two plants of Marseille, the third biggest French town. This activity is also well developed in other countries, particularly in Austria and Central Europe. EDF is now the first cabling operator in the Alsatian region and therefore has a good national base in this market. Investments of FF2 billion are expected over the next ten years, but the public utility ought to limit its commercial developments to small and medium-sized towns.

IV.F. Energy service approach and demand-side management niches

The public utility has for a long time been a technology-orientated company, facing a rapidly growing demand and providing electricity. In the 1980s, which saw the end of the large investment programs, the resulting over-capacity, the saturation of traditional markets and increasing competition between energies have entailed a transition to a more customer-focused policy. The public utility strongly reinforced its strategy by adopting the notion of customers service. The creation of the Directorate for Strategic and Commercial Development in 1988[22] and the new designation of EDF-GDF Services for the Directorate of Distribution directly result from the effort to 'fidelize' existing customers for competitive uses of electricity. Furthermore, the State-owned company adopted a new modern managerial policy, i.e. to gain non-traditional markets, and started to develop national and international activities directed towards a world-wide presence and increasing profits.

The State-owned company has usually been able to propose special deals for the largest customers. This policy can be successful only if such contract terms remain quite opaque, so that other clients cannot bring pressure to negotiate their own deals.[23] The current introduction of competition on the supply side could reinforce this appeal. The utility has also increased the variety of its tariffs for a considerable period. Different cut-off options during peak-load hours enable it to propose lower tariffs in exchange. Regarding domestic customers, large advertising campaigns for electric heating and air-conditioning, or showing a cheap and clean kWh have been promoted. The utility has also offered new services, such as the successful 1994 'service quality insurance' (based on punctuality and improvements in the relationship between the utility staff and the customer), the 'advising service' proposed in 1995, and the new TEMPO tariff option.

The current French experimentation in demand-side management is another example. Even if the French involvement was subsequent to foreign experiences, a national agreement between the Ministry of Industry, the utility and ADEME was signed in January 1993. It could not be a mere adoption of foreign DSM models because of French specificities. DSM is considered to be only a complement to the electricity pricing system, which is usually said to be highly efficient and has been a powerful DSM tool for three decades. Three different geographic areas were defined as DSM niches.

According to the agreement, small testing programs in peri-urban areas mainly concern poorly known captive uses of electricity. FF300 million over a three-year period was available. After two years, results

---

[22] Now called Directorate of Development.
[23] The contract with Péchiney for its aluminium smelter in the North of France is an example.

were real but limited, with only FF21 million spent. The large-scale extension of these actions all over the country is uncertain, and has been deeply criticised by the utility. The second DSM experimental axis is taking place in rural areas, where price equalisation hides the high distribution costs. FF100 million have been allocated by the FACE for testing DSM and renewables in areas such as possible substitutes to extension or reinforcement of the local grid. The third way is being tested in the French overseas territories,[24] where the gap between the average generation cost (1.30 FF/kWh) and the tariff to the customers (0.66 FF) is very wide, inducing losses for the public utility (FF1.8 billion in 1993). DSM in these territories could solve local problems of production, distribution and development, but it must be pointed out that in each area, DSM remains at the experimental stage and its future is still uncertain.

IV.G. *Internationalisation policy of the State-owned company*

During recent years, many events have clearly shown that the State-owned company wishes to remain one of the largest international operators in the power sector. Beyond its traditional technical co-operation with former French colonies in Africa and its huge exports in Europe, its recent industrial interests induced the firm to pay particular attention to international developments.

It is also clear that the company wishes to assume an international leadership position concerning nuclear technology. The agreement between Framatome and Siemens to develop the next generation of reactors, and the building and testing of the Daya Bay and Ling Ao nuclear plants in China point to this direction. Another example is the involvement in eastern European utilities (Mochovce in Slovakia, Rovno in the Ukraine, Bulgaria and Russia) to upgrade nuclear safety in existing and commissioned nuclear power plants, by providing scientific support or assistance.

Abroad, the company has supplied the whole range of electricity engineering services for a long time. More recently, it started to develop a strategy of direct investment in foreign electricity utilities for generation, transmission or distribution. The holding company EDF International SA has bought shares of utilities in Spain, Portugal, Argentina, Ivory Coast, Guinea, Sweden, Brazil, etc. Annual investment increased from FF300 million in 1993, to FF3 billion in 1995, and reached FF5 billion this year.

---

[24] Guadeloupe, Guyana, Martinique and Reunion.

Table VII.7 Selected EDF shares

Country	Activity	EDF share
Generation		
Spain/Elcogas	Investment: combined cycle, natural gas 330 MW	Shareholder of the main investor
Italie/ISE	Electricity generation	With Edison
Portugal	Investment: 2 × 300 MW (coal)	10%
Argentina	Two hydropower plants (670 MW)	51%
Transmission		
Europe	Interconnecting grids with an expansion to eastern grids	
Argentina/Distrocuyo	Operator of high-voltage transmission	36.2%
Other	Consulting engineer in Croatia, Syria, Indonesia, etc.	
Distribution		
Hungary/Edasz et Demasz	Distribution to 1.6 million customers	48%
Argentina/EDENOR	Operation of the EDENOR network in Buenos Aires—estimated FF5.3 billion turnover in 1995	Leadership
Vertically integrated utilities		
Sweden/Sydkraft	1994 sales revenue: FF8 billion	10%
Guinea/SOGEL	Operation of the national grid	In association with SAUR (France)
Ivory Coast/CIE	Operation of the national grid	In association with SAUR (France)
Brazil/Light	Ownership and operation of the local grid	In association with Brazilian and US firms
Other activities	Rehabilitation (Bosnia, Cambodia) electrification (Indonesia, Burkina Faso, etc.)	

The State-owned company does not limit its international diversification strategy to the electricity market. It goes into the large domain of local public services (waste management, water distribution, TV cabling). Waste management appears to be the main field for international as well as national diversification. With TIRU, the company is now active in various countries by operating incineration plants (USA, Canada and Spain) or by purchasing national companies (as ASA, the leader of the Austrian waste management market). To facilitate the conquest of such foreign markets, EDF International set up SAUR International with Bouygues (the world leader in public works). With other industrial partnerships (Total, Elf and Usinor), the company has confirmed its wish to accelerate international diversification in the fields of electricity and local public services. However, it seems contradictory to expect that other countries will allow the French public utility into their electricity sector while, in France, the firm and the Government are defending the public monopoly.

V. Conclusion

Distinguishing developments which might provide the French public monopoly with an unanswerable challenge is not easy. The ambiguities pointed out above will contribute to fix the acceptability of new issues for the electricity regime in the mind of French customers, citizens and politicians. The situation may develop, mainly if liberalisation of electricity sectors in the neighbouring countries turn to be successful.

The system seems stable, but things have already effectively changed. The French organisation could not remain cut-off from the main current developments: the internationalisation of the electricity supply industry, European debate about competition on power generation or third-party access, independent power production development in neighbouring countries and, in France, the new decision-making powers of local authorities, the marginal but efficient competition of large water distribution companies, the deferred dismantling of the oldest nuclear plants (nuclear-plant life has increased from 25 years to 40 years) and the deferred debate about their replacement.

The international challenge is two-fold. On the one hand, the threat of the EC's internal market program exists. It compels the French public utility to end its monopoly situation by providing third-party access to the grid and customer contracts with other suppliers. On the other hand, deregulation in other countries is inviting electricity utilities to expand their activities abroad. Both elements are, of course, closely linked.

With its international and national diversification strategies, the State-owned company is clearly anticipating a possible modification of its monopoly position in the internal electricity market. By gaining new implementations in the field of local public services management, the firm is starting an important evolution. On the other hand, the French water companies are preparing for their entry into the national electricity market. These two strategic moves are complementary: they show that the leading players in urban services and electricity are trying to anticipate a possible deregulation of the French energy market and to take strong positions before the arrival of other players, in France and abroad.

The new flexibility and 'commercialism' of the public utility is not merely a successful attempt to prevent customers from demanding a more liberal electricity policy which would give them access to other electricity suppliers at a lower cost. It is also a sign that cost considerations are beginning to influence the search for customers. The decision to opt for a selective development of air-conditioning is another sign that the utility has stopped looking for customers at any cost, but is concentrating on the most profitable customers. These actions, together with the willingness to offer customer-specific tariffs, are a dent in the public service credo.

The relations between the State and the public utility are recorded in the Contract Plan. This price-cap regulation usually incites the utility to reach productivity gains, but its efficiency is limited by the information asymmetry between the State and the utility. Furthermore, the State is not a monolithic structure. If the Ministry of Industry is the technical decision-maker in fine, some other Ministries are involved in the contract discussions (Budget, Treasury, Economy); their goals are sometimes unconnected with each other, reducing the regulator's perspicacity and power.

Literature

AHEF (1994) *Le Financement de l'Industrie Electrique 1880–1980*, AHEF Conference 1992, PUF, Paris.

AHEF (1986) *La France des electriciens 1880–1980*, AHEF Conference 1985, April 16–18, PUF, Paris.

Allias, M. (1989) *La théorie générale des surplus*, new edition. Presses Universitaires, Grenoble.

Bacher, P. (1993) EDF et le nucléaire en Europe de l'Est, *Revue de l'Energie*, No. 445, January.

Bauby, P. (1994) Electricité et Société. In *Cahiers de Prospective*, Inter Editions, Paris.

Boiteux, M. (1996) Concurrence, régulation, service public—variations au tour du cas de l'électricité, *Futuribles*, January.

Boîteux, M. (1987) Le calcul économique dans l'entreprise électrique, *Revue de l'Energie*, No. 390, February–March.

Bouttes, J. P. and Lederer, P. (1990) Sur l'ouverture des réseaux électriques en Europe, EDF, *Etudes Economiques Générales* M.90.22, March.

Cauret, L. and Adnot, J. (1996) *Why optimize an already efficient system? Overview of the French DSM approach*, 19th IAEE Conference, May 27–30, Budapest.

Colombier, M. (1992) Régulation économique et projet technique: le jeu des compromis entre efficacité, équité et innovation dans le cas de l'électrification rurale en France. Thèse doctorate, Paris, EHESS.

Compagnie Générale des Eaux (1995) *Rapport d'activités 1994*, Paris.

Conseil d'Etat (1995) Service public, services publics: déclin ou renouveau? In *Rapport Public 1994, Etudes et documents no. 46*, pp. 15–134, La Documentation Française, Paris.

Conte, R. (1961) Les origines de l'industrie de l'électricité en France, *Revue Générale de l'Electricité*, No. 1.

Cosnard, D. (1996) Europe de l'électricité: ce qui va changer. In *Les Echos*, Industrie, Paris.

Decré, F. and Chefdeville, H. (1995) Principes de tarification de l'électricité en France. In *Techniques de l'Ingénieur*, chapter traité Génie Electrique D 4 023, EDF, Paris.

Dessus, G. (1971) Les principes généraux de la tarification dans les services publics. In *Vingt-cinq ans d'économie électrique*, Morlat and Bessière, ch. 13, pp. 243–264, Dunod, Paris. First published in *Congrès de Bruxelles de l'Unipede* (1949); published in English in *International Economic Papers*, I (1951).

EDF (1996) *Rapport annuel 1995*, Paris.

EDF (1996) *International activities*, Paris.

EDF (1995) *L'international: la stratégie d'une nécessité*, Sans Frontières, hors série, Paris.

EDF (1995) *Annuaire Statistique 1994*, Paris.

EDF (1995) *Faits marquants 1994, Direction de l'Economie, de la Prospective et de la Stratégie*, Paris.

EDF (1993) *Réussir ensemble notre plan stratégique d'entreprise 1993–1995*, Paris.

EDF (1979) *Le calcul économique et le système électrique*, Eyrolles, Paris.

Fauve, M. (1993) L'Europe de l'Est, un défi pour EDF, *Revue de l'Energie*, No. 445, January.

Finon, D. (1996) Les nouvelles fonctions du régulateur et du gouvernement dans les industries électriques libéralisées: les leçons des expériences européennes, *Revue de l'énergie*, No. 477, April–May.

Finon, D. *et al.* (1995) Changements dans les industries électriques. *Revue de l'Energie*, numéro spécial, No. 465, January–February.

Finon, D. (1993) *La politique énergétique française: efficacité et limites du Colbertisme*, draft, IEPE, Grenoble.

Frost, M. (1985) Economists as nationalised sector managers: reform of the electrical rate structure in France 1946/1969, *Cambridge Journal of Economics*, September.

Gallois, D. (1996) L'Europe de l'électricité est confrontée à la déréglementation mondiale, *Le Monde*, May 7.

Ginocchio, R. (1963) *Législation et organisation de l'industrie électrique*, Eyrolles, Paris.

Gouvello, C., Poppe, M. and Hourcade, J. C. (1995) Maîtrise de la demande d'électricité et surcoûts de la desserte électrique rurale, *Report to the Ministry of Energy and ADEME*, CIRED, Paris.

Henry, C. (1984) La microéconomie comme langage et enjeu de négociations, *Revue Economique*, No. 1, January, pp. 177–197.

Hourcade, J. C. (1991) Calculs économiques et construction sociale des irréversibilités: leçons de l'histoire énergétique récente. In *Les Figures de l'Irréversibilité en Economie*, pp. 279–310, EHESS, Paris.

Kalaydjian, R. (1992) *Le secteur électrique ouest-européen face aux enjeux de l'intégration communautaire*, CIRED, Paris.

Launet, E. (1996) L'Europe libère son énergie, *Libération*, June 21.

Lyonnaise des Eaux (1995) *Rapport d'activités 1994*, Paris.

Martin, J. M. (1993) Les industries électriques: des monopoles contestés, *Economie et Politique de l'Energie*, Collection Cursus Economie, Armand Collin ed., Paris.

Menage, G. and Ailleret, F. (1995) Electricité de France en 1994, *Révue de l'Energie*, No. 466, March, pp. 156–160.

Ministère de l'Industrie (1995) *Gaz Electricité Charbon Rapport 94*, DGEMP, Paris.

Ministère de l'Industrie (1995) *Production–Distribution de l'énergie électrique en France: Statistiques 1994*, DGEMP, Paris.

Ministère de l'Industrie (1993) *Contrat de Plan entre l'Etat et Electricité de France 1993–1996*, Paris.

Ministère de l'Industrie (1993) *Réforme de l'organisation électrique et gazière française* (the Mandil report), DGEMP, Paris.

Ministère de l'Industrie, Ministère de l'Environnement, Ministère de la Recherche (1994) *Débat National Energie & Environnement–Rapport de Synthèse* (the Souviron Report), Paris.

Naudet, G. (1993) Prospective électronucléaire mondiale 2020 dans le contexte énergétique global, *Revue de l'Energie*, No. 448, April.

Persoz, H. and Remondeulaz, J. (1993) La coopération dans le domaine des réseaux électriques entre l'Est et l'Ouest de l'Europe, *Revue de l'Energie*, No. 445, January.

Puiseux, L. and Boiteux, M. (1970) Neutralité tarifaire et entreprises publiques, *Bulletin de l'Institut International d'Administration Publique*, No. 12, pp. 108–121, Berger-Levrault, Nancy.

Ramband-Measson, C., Baumgertner, T., Shanker, A. (1994). *Electricity Regimes and Environmental Policies in France; An Analytical Perspective*, mim., pp. 29, IED, Paris.

Sablière, P. (1993) La loi du 8 avril 1946 commentée, *Cahiers Juridiques de l'Electricité et du Gaz*, hors série, February.

Stoffaës, C. (1994) *Entre monopole et concurrence—régulation de l'énergie en perspective*, PAU, Paris.

Van de Vyver, ?. (1995) *Le service public local: mode d'emploi des Régies et des Concessionnaires*, IVF Services.

Wilsford, D. (1989) Tactical advantages versus administrative heterogeneity: the strengths and the limits of the French State. In James A. Caparano (ed.) *The Elusive State: International and Comparative Perspectives*, pp. 128–172, Sage, London.

Chapter VIII
The German Electricity Reform Attempts: Reforming Co-optive Networks

LUTZ MEZ

The electricity industry belongs to the most powerful companies in Germany, and constitutes an economic and political power cartel which until today was able to avoid all attempts to change the framework conditions for energy policy in Germany. This power position is based on the one hand on the economic strength of this sector. The growing importance of electricity supply in Germany is reflected by a 25% increase in electricity demand in the period 1978–1989, while final energy use declined by 10%. For decades, public utilities have been the biggest investors, controlling the energy system from the open-cast mining of lignite to the generation and distribution of electricity to the last light bulb. In 1994, the public utilities invested DM14.1 milliard and employed 196,315 persons. On the other hand, structural defects in the state's energy policy in the form of a 'state failure' (Jänicke, 1990, pp. 58ff) in the control of the electricity industry, and the entanglement of state bodies with the industry at all levels, strengthen the industry's position in Germany.

I. Historical Tradition

The beginning of the electricity industry in Germany was in the period between 1880 and 1890. The first public power stations were started in Stuttgart in 1882, followed by Berlin in 1884, Dessau and Lübeck in 1886, and so on. The building of city power stations for electrical trams and the founding of city-owned utilities continued until 1900, and then the period of long-distance power stations ('Überlandzentralen') supplying the countryside followed. During World War I, large power stations were built. The national grid and its controlling companies ('Verbundwirtschaft') have dominated development since the 1930s (Zängl, 1989, pp. 9ff). The historical and political process is characterised by the attempts of the electricity supply industry to continually increase electricity consumption and by the predominant non-intervention of the State (Zängl, 1989, p. 7).

After World War II, the electricity industry in West Germany remained nearly untouched by the Allies, and the electrification of the country continued. From the 1960s, nuclear power stations came into operation, and the profile of the industry was, until 1986, imprinted by the nuclear program. Since the Chernobyl catastrophe, the phasing out of nuclear power has been on the political agenda, not only in the form of the manifesto of the Green party and environmentalists, but also as a policy of the Federation of German Trade Unions (DGB) and the Social Democratic Party (SPD) to phase out nuclear power plants in Germany within the next ten years.

As early as 1976, the German Monopolies Commission had criticised the electricity supply industry as belonging to 'economic sectors without competition' (Monopolkommission, 1976, pp. 52, 382ff). This was not always the situation. Before World War II more than 16,000 public electric utilities existed in Germany. In the 1950s there were about 3500 in the Federal Republic of Germany alone. Today there are about 900.[1] This shrinking process in no way explains the real power concentration on the electricity market.

II. The political economy

On the level of the national grid constitutes the nine companies shown in Table VIII.1. All of the companies of the 'Verbundwirtschaft' are interconnected through capital links (Fig. VIII.1). They are joint members of the association 'Deutsche Verbundgesellschaft eV' (DVG). The DVG was founded in 1948, before the Federal Republic of Germany was constituted. Decisions concerning the national grid and projections of new power plants are made initially by the DVG. The DVG members dispose of 91,152 MW (75%) of the total installed capacity of power stations. The net electricity supply of the DVG members in 1994 was 366.4 TWh. Of this, 90.1 TWh was delivered to special contractors

---

[1] In 1994 the public electricity supply in Germany comprised 706 utilities, 659 in West Germany and 47 in East Germany. An additional 110 new municipal utilities were founded in the meantime in East Germany.

Table VIII.1 National grid companies in Germany and their own plus contracted power capacity (MW) in 1994

Badenwerk AG	5034
Bayernwerk AG	8693
Berliner Kraft- und Licht (Bewag) AG	2847
Energie-Versorgung Schwaben AG (EVS)	4666
Hamburgische Electricitäts-Werke AG (HEW)	3823
PreussenElektra AG	18,597
RWE Energie AG	25,960
Vereinigte Elektrizitätswerke Westfalen AG (VEW)	6719
Vereinigte Energiewerke AG (VEAG)	14,813
Total	91,152

Source: VDEW (1995, p. 19).

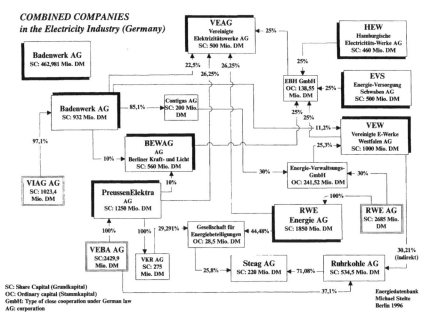

Fig. VIII.1. Capital links among German utilities. *Source:* Stelte (1994).

(Sondervertragskunden), 41.3 TWh went to normal tariff customers (Tarifkunden) and 235.1 TWh was taken over by other utilities for further supply (Table VIII.2).

Table VIII.2. Electricity supply of DVG members by customer groups, 1994

	Total supplied (TWh)	Electricity supplied to					
		Special contractors		Tariff customers		Indirect supply	
		TWh	%	TWh	%	TWh	%
Badenwerk	19.4	4.6	23.5	4.0	20.5	10.8	56.0
Bayernwerk	36.2	3.3	9.1	0	0	32.9	90.9
Bewag	13.2	6.1	46.0	7.1	54.0	0	0
EVS	19.9	4.9	24.7	4.8	23.4	10.2	51.0
HEW	12.4	8.5	68.6	3.9	31.2	—	0.2
PreussenElektra	58.3	4.7	8.0	0	0	53.6	92.0
RWE Energie	123.6	46.7	37.8	14.5	11.7	62.4	50.5
VEAG	50.1	0.1	0.1	—	0.003	50.0	99.9
VEW	33.3.	11.2	33,6	7.0	20.9	15.2	45.5
DVG	366.4	90.1	24.6	41.3	11.3	235.1	64.1

Source: DVG-Jahresbericht 1994, FFU calculation.

II.A. Electricity generation

German electricity generation in 1994 was undertaken by three sectors: public utilities (455.5 TWh), industry and coal-mining companies (65.1 TWh), railway companies (6.2 TWh). Most public electricity generation takes place on three levels:

1. by the nine national grid companies (Verbundwirtschaft) (*c.* 366 TWh);
2. by the 63 regional companies on a regional level (*c.* 38 TWh); and
3. by *c.* 570 city and town utilities at a local level (*c.* 40 TWh)

The Verbundwirtschaft generally covers the base-load supply while the regional and local companies cover the medium- and peak-load markets.

The national grid companies are mainly wholesale producers, with about 65% of their production sold to other electricity companies. Only *c.* 10% of their production is sold to tariff customers. The regional and local companies are mainly distributors with, respectively, 80 and 63% of their sale bought from the Verbundwirtschaft.

A comparison of electricity supply and companies clearly demonstrates the differences in the size of the DVG member companies. RWE alone generates more than a third of the total electricity, whereas Bewag only accounts for 3.6%. The data on different sectors of consumption shows the alternating importance of special contractors. HEW delivers almost 69% of its electricity to these, while this figure for VEAG is only 0.1%. Also visible from Table VIII.2 the different structure in respect to regional supply companies and local utilities (Stadtwerke). VEAG, Bayernwerk and PreussenElektra do not supply tariff customers directly, but through regional companies. For HEW and Bewag, in contrast, indirect delivery is of no importance because both are also local utilities.

At the regional level, electricity supply is carried out by regional companies. They work together in the Association of Regional Energy Supply Companies (ARE), whose membership numbers 53 companies. An additional ten non-associated companies exist, so the number of regional companies grows to 63. Of these companies, 57 are directly or indirectly governed or strongly influenced by the big utilities. Only six regional utilities are independent of capital links with the Verbundwirtschaft.

On the local level, municipal utilities (Stadt- und Gemeindewerke) are involved in the German electricity market. The Association of Local Companies (VKU) had 488 companies engaged in electricity supply in 1995. An analysis of West German utilities shows that the big utilities in 1993 owned shares of between 10 and 99% in 57 of the local utilities (see Fig. VIII.2). The relatively large number of electricity companies in Germany compared to other EC countries, has to be seen in the light

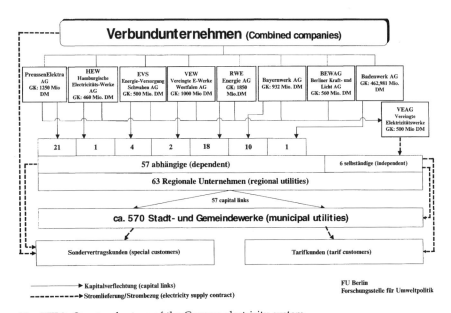

Fig. VIII.2. Structural set-up of the German electricity system.

of double control, partly through strong capital linkage and partly through electricity supply.

The Verbundwirtschaft is also the main electricity generator. The total electricity generated in Germany amounted to 526.8 TWh (excluding consumption by power stations) in 1994. Of this amount, about 455.5 TWh (86.5%) was produced by public utilities. Electricity generated by the Verbundwirtschaft accounted for 366 TWh, so that 80.4% of electricity produced by public utilities comes from only nine companies. The share of the Verbundwirtschaft of total generation in Germany was 69.6%.

The Regional Supply Companies (only members of the ARE) directly supply 33 million customers and indirectly a further 12 million. Electricity supply amounted to 196.5 TWh in 1995, which constituted a share of more than 40% of the total supply from public utilities. In this category also, major differences in size exist. The smallest company delivers only 0.4 TWh, and Energieversorgung Ostbayern, which is the largest regional company, over 11.5 TWh. While ARE-member companies have a major share of electricity supply, their role in energy generation is limited; only 17% of total supply is self-generated, and 81% is contracted from the Verbundwirtschaft. The remaining 2% is due to internal transfers. The domination of the Verbundwirtschaft is apparent here. The small share of the ARE correlates with their self-image of being primarily responsible for energy distribution across the country. Following the restructuring of the electricity sector in eastern Germany, the regional companies' share of energy supply has decreased further. The regional companies of the new German Länder own only limited generation capacities, and are provided with 95% of their electricity by the VEAG. The limited role of electricity generation for the regional companies is further highlighted when electricity supply and power-plant capacity are compared. The maximum capacity of ARE power plants in 1994 was 9157 MW. As 34.5 TWh was produced, the average utilisation time was 3773 hours. Power plants of the Verbundwirtschaft are, on average, in use for 4600 hours per year. The Verbundwirtschaft therefore primarily covers base-load demand, whereas regional companies operate in the medium- and peak-load markets. This is a major factor for the higher generation costs incurred by regional companies. On the other hand, regional utilities dominate city and town companies with respect to electricity supply. They transport power to 554 local distribution companies, and thereby secure the supply of *c.* 12 million people (ARE, 1996, p. 31).

Local utilities are in a similar situation to regional utilities. They are primarily active in the field of distributing electricity generated by regional or national grid utilities. From 1961 to 1990 the share of power being self-generated dropped from 35% to only 20%. Member companies of the VKU of the West German Länder supplied 130 TWh in 1993, which counts for about 30% of the total public supply of electricity. Of the 491 VKU members active in the electricity market, 170 companies have no generation facilities of their own, and a further 321 companies generates a share of 0–30% of their supply. Only 41 local utilities (9.2%) generate more than 30% of the power they distribute. At the same time, 75% of power-plant capacity is in the ownership of only 10 companies, such as the Technische Werke Stuttgart and the Stadtwerke of Hannover, Bremen, Kassel, and Saarbrücken. The installed generation capacity of local utilities amounted to 11,120 MW in 1993, which constitutes a share of 9.1%. They were producing 39.9 TWh and providing 8.8% of the public supply. According to the VKU, the gap between capacity and actual supply comes from the fact that more efficient base-load supply is provided by national grid utilities, and that CHP has a higher priority in local utilities. In general, local utilities cover the medium- and peak-load demand.

Another important sector in the electricity market is that of the industrial producers. Primarily, they are responsible for generating electricity for industry and the railways, but they transmit their surplus capacity to the grid. Generation from the production sector amounted to 36.6 TWh in 1994, representing a share of 6.9% of electricity in Germany.

A special role in the German electricity market is played by the coal-mining companies. STEAG (owned by Ruhrkohle AG, share of 71.08%, the Gesellschaft für Energiebeteiligung, share 25%, and the RWE-DEA AG, share 1.62%) and Saarbergwerke (owned by the Federal government, 74% share, and the Saarland, 26% share) together own a formidable capacity of about 6000 MW. In 1996, Ruhrkohle and Saarbergwerke merged. They supply electricity for the public grid, even though they do not belong to the national grid utilities. Together with the capacity of the lignite-mining companies, the mining sector generated 28.5 TWh of electricity in 1994.

The electricity activities of the mining and production sector are often merged in the statistics. Their organisation, the Vereinigung Industrieller Kraftwirtschaft (VIK), represents about 90% of all the participants. Although the share of independent industrial generation has risen in recent years, the volume of electricity produced is still far below the amount of the early 1970s, when about 70% of industrial demand came from industry's own installations. According to the National Statistical Office, industrial utilities today have an installed capacity of 15,175 MW, of which 3025 MW are in the new Länder. Since 1970, the volume bought from public utilities has grown faster than total demand.

One reason for this is that a number of power stations fell statistically into the sector of public utilities. The generation capacity of VEBA, of Klöckner, Thyssen and Volkswagen were taken over by national utilities. The main reason for the decrease in industrial capacity is thought to be the instrument of special contracts which convinced industry to resume its generation in favour of the national utilities.

A new tendency in the industrial sector is the contracting of not only small-scale power plants, but also large CHP plants. The chemical company BASF, the largest single energy customer in Germany, is contracting a 400 MW combined-cycle power plant from RWE. A similar contract for a 800 MW combined-cycle plant was signed in 1995 between RWE and the chemical company Hoechst.

Of more statistical relevance is generation by Deutsche Bahn AG, the railway company. Installed capacity amounted to 1523 MW in 1994, the volume generated being 6.2 TWh.

II.B. Electricity supply to special contractors and to tariff customers

The purchase of electricity by special contractors may be negotiated under special conditions which are not subject to the General Conditions of Supply ('Allgemeine Versorgungsbedingungen', or 'AVB') or the Federal Tariff Regulation on Electricity ('Bundestarifordnung Elektrizität', or 'BTO Elt'). They are also exempt from national control of prices. Special agreements apply to large industry customers and public facilities. The supply of special contractors is practised on all the three levels—national, regional and local. According to Table VIII.3, between 6–25% of total electricity supply is consumed by special contractors. In aggregate terms, 245.9 TWh, which is 55% of total net consumption, is used by special customers.

The supply of tariff customers is secondary to the supply of those with special contracts. Tariff customers are those whose prices are

Table VIII.3. Total electricity supply and supply for special contractors, 1994

	Total supply (TWh)	Supply to special contractors	
		TWh	%
National grid utilities	366.4	90.1	24.6
Regional utilities	193.0	74.9	38.8
Local utilities	130.7	80.9	61.9
Total	447.2	245.9	55.0

set according to the General Conditions of Supply ('Allgemeine Versorgungsbestimmungen', or 'AVB') and the National Tariff Regulation for Electricity (Bundestarifordnung Elektrizität). Of the total German electricity consumption of 447.2 TWh, 172.8 TWh (38.6%) was used by tariff customers in 1994. Together, they do not create a homogenous group. Parts of industry and commerce, public facilities, agriculture, and primarily households are included. Domestic supply alone is 124.5 TWh.

II.C. Electricity transmission

Among investments in the German electricity sector, national transmission and local distribution are of growing importance. According to the VDEW, the total length of all transmission systems amounted to 1.5 million km in 1994, of which 1.235 million km were in the western Länder. The grid is to be divided into a high-voltage network of 110–380 kV, a medium-voltage network of up to 60 kV, and a low-voltage grid of under 1 kV. As Table VII.4 shows, the three-dimensional structure of the German electricity sector is also mirrored by the availability of the grid. The high-voltage grid is predominantly in the hands of the national grid utilities. Of the almost 115,000 km of the high-voltage network, about 73,000 km (64.4%) is owned by the nine big utilities. Regional utilities own about 35,000 km (30.1%), and the remaining 6000 km is shared by industry and local utilities. These facilities serve national and international transportation and the supply of demand centres. In contrast, the medium-voltage grid serves the purpose of

Table VIII.4. Ownership of the German electricity grid, 1994 (in km)

	Under 60 kV	110–380 kV	Total
Badenwerk	—	4022	4022
Bayernwerk	—	5614	5614
BEWAG	—	523	523
EVS	—	5416	5241
HEW	—	1385	1385
PreussenElektra	—	18,385	18,385
RWE	—	21,063	21,063
VEAG	—	11,300	11,300
VEW	—	5345	5345
DVG	—	73,053	73,053
ARE	764,644	35,393	800,037
public supply	1,399,870	114,981	1,514,851

further transportation, and has a length of 466,463 km. Accordingly, the share of regional utilities is higher (above 56%). Unfortunately, national grid utilities (with the exception of Badenwerk) do not publish any official figures on this matter. It can be presumed that the remaining transmission lines are in the hands of the national utilities and local companies. About 926,000 km of the low-voltage grid, which is responsible for distribution to the small consumers, is predominantly owned by regional utilities (54.4%) and local utilities.

D. Production characteristics

In 1994 the total installed capacity of public utilities was 104,826 MW, of which 89,289 MW was in the western part and 15,537 MW was in the eastern part of Germany (Bundesministerium für Wirtschaft, 1996, p. 47). Examining the power-station type and electricity generation, the leading role of nuclear power and lignite stations in public supply can be seen, followed by hard-coal fired power stations. Natural-gas fired stations and hydropower have a share of about only 4% of generation (Table VIII.5). The average utilisation time of nuclear power stations was 6273 hours in 1994 (Bundesministerium für Wirtschaft, 1996, p. 44). Lignite power stations ran for 5914 hours, hydro for 5818 hours and hard-coal fired power stations for 4598 hours. Natural-gas and oil-fired power stations were on average used for 1421 and 478 hours, respectively.

The exchange of electricity across the German border by public utilities in 1994, according to DVG statistics, resulted in 35,724 GWh imported and 33,506 GWh exported, an import balance of 2218 GWh.

Table VIII.5. Installed capacity and electricity generation of German public utilities, 1992

	Capacity (MW)	Generation (GWh)	Share (%)
Lignite	22,207	136,065	29.9
Hard coal and mixed fuel	26,760	118,348	26.0
Oil	8458	4883	1.1
Natural gas	14,425	23,299	5.1
Nuclear	23,770	149,983	32.9
Hydro	8351	20,129	4.4
Others	853	2841	0.6
Total	104,826	460,823	100.0

Source: Bundesministerium für Wirtschaft (1996, pp. 44, 47) and FFU calculation.

Table VIII.6. Import and export of electricity by German public utilities, 1994

Country	Import (GWh)	Export (GWh)	Balance (GWh)
Austria	4749	4855	−106
Switzerland	6553	6736	−183
France	15,333	379	+14,954
Luxembourg	423	3697	−3274
Netherlands	679	13,037	−12,358
Denmark	3601	175	+3426
Sweden	29	40	−11
Poland	3180	4411	−1231
Czech Republic	1177	176	+1001
Total	35,724	33,506	+2218

Source: BMWi (1996, p. 55).

Table VIII.6 shows electricity exchange values for neighbouring countries.

III. Legal Foundations of the German Electricity Sector

The legal framework for the German electricity sector consists of a mixture of public and private laws. A wide range of public laws, directives and guidelines, as well as legal rulings and private contracts, constitute the regulatory framework of the electricity sector.

The legal basis of the German electricity sector is the Law for the Promotion of Energy of 1935[2] ('Energiewirtschaftsgesetz', or ' EnWG'), which should have been amended during the last legislative period. According to the EnWG, which once served as a preparation for war, the supply of electricity should be as economical and safe as possible. Preservation of resources and environmental protection are to be included as aims of equal importance. As a balance to the general obligation to supply, which is in effect often waived, the public utilities act within a monopoly market. Other utilities or companies are only allowed to operate after positive decisions by the regulatory authorities, and after they have informed the public utilities. On this basis, a system has developed which is organised through demarcation treaties between suppliers, and concession treaties between suppliers and local authorities. Through contracts of demarcation, public utilities secure their territories on a private-law basis, avoiding competition of supply. Such

---

[2] This law is also called the 'Energy Management Act'.

contracts exist on local, regional and national levels, and originated from conflicts which have been running since the 1920s. Through contracts of concession, utilities are granted right of way by the local authorities. In return, they pay concession tariffs to the municipality. Such contracts are signed on the basis of the Directive on Concession Tariffs ('Konzessionsabgabenverordnung', or 'KAVO') of 1941. Until 1991, the concession tariff was set at a range of 10–18% of the sale of gas and electricity to consumers. Since reform in 1991, the amount has been calculated on the basis of kWh supplied. For local authorities, concession tariffs are an important source of income, which explains their interest in high sales of gas and electricity.

The Law against Limitation of Competition ('Gesetz gegen Wettbewerbsbeschränkungen', or 'GWB') exempts contracts of demarcation from the general prohibition of cartels. The fourth amendment to the GWB has tried to introduce limited competition, as it has set a period of 20 years as the maximum duration of concession treaties. Until 1994, all existing concessions had to fulfil this condition. The fifth GWB amendment of 1989 additionally established the synchronisation of the duration of concession and demarcation treaties. After concessions have expired, a change of supply is possible regardless of any existing demarcation contracts.

Of fundamental importance is the difference between tariff customers and special contractors ('Sondervertragskunden'). Special contractors receive electricity on special conditions—their demand is not calculated on a tariff basis. Tariff structures are regulated through the Federal Tariff Order for Electricity (Bundestarifordnung Elektrizität, BTO Elt) and for gas (BTO Gas). Only electricity tariffs are subject to official control.

Of growing importance with respect to the electricity sector are environmental laws. The Law against the Pollution of Air ('Bundesimmissionsschutzgesetz', or 'BImSchG') was enacted in 1974. The Federal Government enacted the Ordinance on Large Combustion Plants ('Großfeuerungsanlagenverordnung', or 'GFAVO') in 1983 in order to reduce emissions at source. The GFAVO sets strict limitations for all emission components, such as sulphur dioxide, nitrogen oxides, carbon monoxide, halogen compounds and particles. The requirements laid down are categorised on the basis of thermal output rating classes, and for existing plants by setting an explicit timetable for retro-fitting. Plants which had not been retro-fitted had to be shut down by the end of 1993. After the unification of Germany, a similar timetable was set up for power plants in eastern Germany. Retro-fitting of fossil-fuel power stations until 1996 or a shut down after 10,000 or 30,000 hours had to be chosen by the utilities.

With its limitation of private property rights in favour of the environment, the GFAVO constitutes a positive development in regulatory instruments. For the first time, emission limits were not only limited to newly built power stations, but also included existing ones. The retrofitting and shut-down of existing plants was considered necessary within certain time periods. The Technical Instructions on Air Quality Control ('Technische Anleitung Luft', or 'TA Luft') were supplemented in 1986 in that concepts of retro-fitting were extended to all facilities which were to be officially sanctioned.

The government of the GDR and the Treuhandanstalt agreed in August 1990 on the so-called 'Stromvertrag' (electricity contract), for the takeover of the East German electricity sector by the big western utilities. The Stromvertrag consists of

1. a contract regarding the takeover of the GDR grid and the national utility by West German national grid utilities; and
2. contracts regarding the 15 regional GDR utilities.

The agreement was aimed at the privatisation of the GDR utilities, with major participation from the West German utilities. The Stromvertrag transforms the existing controversial structure of the West German electricity sector, into the new German Länder. A protest from 164 local authorities at the German Constitutional Court in Karlsruhe against the Stromvertrag was settled with a compromise in the summer of 1993. Nevertheless, the big utilities were still able to gain control over generation, transmission and the lion's share of local distribution in East Germany. They own all the shares of VEAG Vereinigte Energiewerke AG, and a differing majority of the regional utilities. Through joint ventures, they also try to influence the new local utilities.

The Electricity Feed-in Law ('Stromeinspeisegesetz') fundamentally improved the economics of renewable sources of energy. For electricity from renewable sources supplied to the public grid, a minimum remuneration is guaranteed. The amount is far above the amount agreed upon between public utilities and industrial electricity generators in the so-called 'Verbändevereinbarung' for feeding electricity into the public grid.

To support the utilisation of domestic coal in the electricity sector the so-called 'Jahrhundertvertrag' (contract of a century) was in operation until 1995. The mining companies supplied utilities until this time with 40.9 million tonnes of coal annually. All customers had to finance this through a levy, the so-called 'Kohlepfennig', which was collected by the Federal Government. The volume of the levy was calculated from the difference in cost between domestic and imported coal. In 1994, German hard coal was produced for DM290 per tonne on average, while the price for imported coal was DM70 per tonne. According to a calculation made by the OECD in 1992, the various Government and non-Government aids to German coal amounted to DM187.50 per tonne, or a total of DM11.9 billion. Additional assistance, not benefitting current production, figured with further DM12 billion. The total amount of this aid to coal is about DM24 billion annually.[3] After the Jahrhundertvertrag expired in 1995, the level of price support was partly reduced.

An initiative of top managers from RWE and VEBA, together with the SPD and the Prime Minister of Lower Saxony, Gerhard Schröder, to overcome the blocked (nuclear) energy policy situation with 'consensus talks', failed in the autumn of 1993. Instead of an energy consensus, the Federal Government initiated the so-called 'Article Law', which changed certain articles in different energy-related laws. On July 19, 1994, the 'Article Law' was passed by Parliament.[4] Article 1 regulates the subsidy of indigenous hard coal for electricity generation in the decade 1996–2005. For 1996, a subsidy of DM7.5 billion will be given. From 1997 to 2000, hard-coal mining will be subsidised by DM7 billion annually.[5]

The Article Law also introduced direct storage of used nuclear fuel equal to reprocessing as a precondition for operating licences for nuclear power plants, and asked for higher standards for new nuclear power plants. Further changes concerned the electricity feed-in law for renewable energy. Public utilities have to pay 80% instead of 75% of the average price[6] for electricity generated in small hydro power-plants or plants using sewage, landfill gas or biomass.

IV. Politics—Attempts at Reform

The concentration process of the German electrical power industry has led to a high degree of centralisation. In this process, the existing political and economical power structure—and in the past also economies of scale—played an important role. The legal and institutional framework of the electrical power industry cements this structure and secures

---

[3] Cf. IEA Estimates of Government and Non-Government Aids to German Coal Producers. In: *OECD/IEA: Energy Policies of IEA Countries*, 1992 Review, p. ?, Paris 1993.

[4] Gesetz zur Sicherung des Einsatzes von Steinkohle in der Verstromung und zur Änderung des Atomgesetzes und des Stromeinspeisungsgesetzes. In: *Bundesgesetzblatt*, Teil I, Jg. 1994, pp. 1618–1623.

[5] Generation from hard coal will be reduced. If the delta between import and indigenous coal is calculated with DM200 per tonne, about 32 million tonnes coal can be produced per year.

[6] The average proceeds from the sales of power by public utilities in 1994 was 19.12 Pf/kWh.

the privileges of big utilities. The Monopoly Commission criticised this situation in the middle of the 1970s and recommended 'more competition' and more state intervention in the electricity supply industry. Since the late 1980s a new energy policy, for example as spelled out by the Enquete Commission for Climate Protection, calls for electricity savings.

Looking at the energy policy arena, the main power game was played by the Federal Ministry of Environment, backed up by Länder Governments with a stated CO_2 reduction policy, the majority of the Enquete Commission for Climate Protection, and by environmentalists both inside and outside the Bundestag. The opposing forces are represented by the two biggest national grid companies, RWE and PreussenElektra, acting for the electricity industry. The Federal Ministry of Economy, which is responsible for energy policy in general, acts on the side of the big utilities. Concerning deregulation, the electricity supply industry has to oppose the Federal Monopoly Agency ('Bundeskartellamt'), some reformers within the Federal Ministry of Economy, and reform forces at the European level.

Attempts to reform the co-optive networks of the ESI have been numerous. In the middle of the 1980s a strategy for an about-turn in energy policy ('Energiewende') combined with the re-communalisation of energy supply, was formulated by the Öko Institute (Hennicke, 1985). This concept stressed the role of municipalities as energy-policy decision-makers and organising municipal utilities as energy-service companies. On the basis of energy master-plans, local utilities should operate the energy grids. If the grids are not in communal ownership, buying back supply grids from big utilities should be practised. The concept suggested phasing out nuclear power plants, the 'de-cartelisation' of the electricity supply industry, and a strengthening of statuary energy policy intervention.

This strategy was broadly discussed in West Germany after the Chernobyl catastrophe, and influenced not only the SPD and the Greens, but also activists at a local level. In many cities and communities so-called 'energy turnabout' committees were formed. After unification, the establishment of local utilities in the new Länder gave way to a strong municipal interest in energy supply.

The position of the SPD concerning the reform of the ESI was formulated in a draft law for the EnWG in 1991.[7] The basic points of the SPD's position are:

1. 'a re-communalisation' of electricity, gas and district heating supply on voluntary basis;

---

[7] PD-Fraktion BT-Drucksache 12/1490 (1991).

2. the redefinition of energy supply towards energy services;
3. the monopoly position of public utilities remains;
4. 'more competition in energy supply' is too simple a formula;
5. a drastic enlargement of control instruments, e.g. investment control as a licensing procedure with least-cost planning;
6. a general feed-in right in public grids with price guarantee for producers of CHP, renewable energy and waste energy, oriented towards long-term avoidable costs; and
7. a broad palette of state intervention rights to fulfil these aims.

The concept of the Green Party was, to some extent, similar:[8]

1. 're-communalisation' of electricity, gas and district heating supply by municipal ownership;
2. a transition of the utilities from energy supply to energy service companies; and
3. a notable increase in state intervention and control instruments.

In contrast to the SPD, the Green Party will phase out nuclear power stations within one year. In 1996, the Green Party began a new attempt to replace the old EnWG completely. Their aim is to restructure the cable-dependent energy supply industry in Germany in such a way that:

1. climate protection and the conservation of the natural foundations of life are transformed into a commercial profit-making principle;
2. private capital is mobilised for an ecological turnabout in the energy industry;
3. monopolies are deregulated;
4. the energy supply industry is decentralised; and
5. new suppliers are afforded market access.

The Green Party wish to abolish the monopoly structure and separate energy generation, energy transportation and energy distribution in terms of organisation, bookkeeping and ownership.[9]

The PDS/Left List strives for an 'ecological socialisation' and democratisation of the German ESI, and sketched out a different position:[10]

1. the 're-communalisation' of electricity, gas and district heating supply by municipal ownership, unbundling and 'de-cartelisation' of big utilities;

---

[8] Fraktion Die Grünen BT-Drucksache 11/6484 (1990).
[9] Draft law by Michaele Hustedt *et al.* (1996) *Entwurf eines Gesetzes zur Neuordnung der Energiewirtschaft (EnergieG).*
[10] PDS/Linke Liste BT-Drucksache 12/1294 (1991).

2. decision-making by public town-hall meetings, sub-regional planning councils, and independent energy-control commissions; and
3. a considerable increase of control instruments.

On the other hand, the coalition of the the CDU, the CSU and the FDP spelled out an energy policy using the terms 'liberalisation' and 'deregulation' very clearly. In October 1987, the German Bundestag decided to establish the Enquete Commission on 'Preventive Measures to Protect the Earth's Atmosphere'. Study commissions of the Bundestag, a mixed commission of Members of Parliament and experts, are a specific form of parliamentary policy deliberations. According to the recommendations made by the Enquete Commission, CO_2 emissions have to be reduced in Germany by at least 30% by the year 2005 to achieve climate protection. By the middle of the coming century, a reduction of 80% in Germany and a global cut of 50% are considered necessary.[11] The target reductions demanded have been reaffirmed by the follow-up commission of the 12th Bundestag. Suggestions for a new energy policy, with electricity saving as new target, were brought into public discussion by the Enquete Commissions.

A change in the framework conditions of energy policy is emerging, concerning the deregulation of the energy economy and further development of an environmental regime.

The deregulation of the German electricity supply industry has been a political issue since the Federal Government instituted the Deregulation Commission in 1988. The Commission published two reports on six industries in March 1990 and March 1991. The second report includes a chapter on the electricity supply industry (Deregulierungskommission, 1993).

The Deregulation Commission stated in its report that the specific characteristics of electricity do not justify special exceptions, and proposed far-reaching suggestions for the deregulation of the electricity industry. The report criticises as 'undesirable developments' the excessive price of electricity in Germany, concerning the level and structure. Costs and risks can be off-loaded to tariff customers. By exclusion of competition and concession tariffs and the tariff structure for tariff customers, over-high electricity prices can be demanded. Further, the demarcation treaty system prevents the purchase of low-price electricity. The investment regulation practice as profitability control has led to capital intensive electricity supply. Price regulation is reduced

---

[11] *3rd Report of the Enquete Commission: Schutz der Erde. Eine Bestandsaufnahme mit Vorschlägen zu einer neuen Energiepolitik*, Deutscher Bundestag, Referat Öffentlichkeitsarbeit, Bonn (1990).

through same interests of controlled and controller in selling more elec
tricity. Last but not least the subvention of hard-coal raise the price of
electricity.

The Deregulation Commission suggested the prohibition of
agreements such as demarcation and co-operation treaties between utilities, and the exclusive clause in concession treaties with municipalities.
Grid operators should be liable for the duty of the transit of electricity,
and the high-tension grid should be operated as an independent service.
Further, the Commission suggested inviting tenders for the supply of
tariff customers in local grids. These licences should be limited in time,
and regulated by the State. For tariff customers, the introduction of
linear and load-related tariffs was recommended. Concession tariffs
should be abolished step by step, but utilities have to recompense the
municipalities for the costs of using the grid. The integrated supply
system ('Querverbund') for electricity, gas, district heating, water,
public transport, waste disposal and other communal duties should also
be examined. Preconditions for competition can be created by the abolition of regulations on limited competition and by re-regulation of state
interventions in special market segments. State intervention according
to the basic energy law which steers the market, such as investment
control, supply and licensing control, should be abolished. Hard-coal
policy has to be reformed. Protectionism of indigenous coal has to be
brought into line with safe supply only, and can be cut down. Subsidies
for social and regional reasons must not limit competition.

The Federal Government has set the priority for environmental and
energy policy in the 1990s as the formulation and implementation of a
comprehensive strategy towards climate protection. In the summer of
1990, the inter-ministerial working group 'CO$_2$ Reduction' (IMA) was
put in charge of suggestions to achieve the target of reducing CO$_2$
emissions by 25–30% of their 1987 levels by the year 2005. The extent
of CO$_2$ reduction demands energy savings and rational use of energy
on the one hand, and on the other the substitution of fossil energy.

In this context, the Federal Government has taken the decision to
reform the EnWG by introducing competition and environmental protection. The potential for deregulation will be used as far as possible to
strengthen the scope of enterprise. Environmental protection and careful
use of resources, as well as from the security and low pricing of energy
supply, will be of equal priority in the decision-making of State
energy control.

In November 1989, the Bundestag asked the Federal Government to
present a review of the EnWG, with the proviso that environmental
protection and careful use of resources should be of equal importance
to the target of a cheap and safe electricity supply. The cabinet decision

of November 7, 1990 on the reduction of CO_2 emissions took this up, and put the BMWi (Bundesministerium für Wirtschaft) in charge of presenting a new version of the EnWG. The coalition treaty of the Conservative–Liberal Government also states that the EnWG will soon be amended, and that environmental protection has to be included in the target catalogue of the law.

In March 1992, the Federal Ministry for the Environment (BMU) presented a list of the defects of the EnWG from an environmental perspective (Bundesministerium für Umwelt, Naturschutz und Reaktorsicherheit, 1992):

1. insufficient consideration of environmental concern during investment;
2. energy efficiency of generation;
3. combined heat and power (CHP) in connection with the placement of power stations;
4. energy saving instead of capacity expansion;
5. insufficient usage of the existing or potential electricity generation capacities of self-use producers;
6. insufficient incentives for energy saving through tariff and price structures;
7. missing instruments for the limitation of electricity demand and for the development of district heating;
8. weakness of energy policy regulation instruments, and its use for technical, economical and ecological complexity;
9. information lacks on the side of the control bodies *versus* utilities; and
10. personally, financially and politically limited possibilities of the administration for control and sanctions.

Concerning the introduction of a CO_2/energy tax, the Federal Government welcomes the targets of the EC Commission's initiative to develop a comprehensive European strategy in order to reduce CO_2 emissions through a combined CO_2/energy tax for the internalisation of external effects through energy generation. Further clarification is needed of the basis for calculation, the yield, the time-frame and the concrete shaping of the tax. Mainly the BMU argued for the introduction of environmental charges and taxes in order to encourage ecological governance of competition processes in electricity markets. The introduction of a combined CO_2/energy tax will follow legal directives from the European level. Further fees for grid connection and use of electrical equipment will be developed as instruments for environment-related market intervention. The BMU concept aims to connect competition in the electricity supply industry with market intervention through environmental charges and taxes and State control of electrical grids.

In September 1993, the BMU published a concept for reforming the ESI, which consists of three elements:

1. functional unbundling of electricity generation, transport and supply to electricity end-users by introducing competition in the generation and supply sector, and reducing State control on electrical grids;
2. organisation of electricity markets by the creation of pools; and
3. environment-oriented market intervention through environmental charges and taxes, and fees for grid connection and usage.

A draft by the BMWi for the review of the German energy supply law (EnWG) was circulated from October 1993 with the aim of encouraging deregulation and competition in gas and electricity markets in order to lower energy prices. The draft formulated a special form of unbundling. Energy supply companies of a certain size should render accounts for electricity generation and distribution separately. Third-party access will be possible, because the owner of the grid has to prove why trespassing is not possible. State supervision of the ESI will be reduced to essential measures, for example, control of the price of electricity. Investment control for power stations and grids are, according to the BMWi, no longer necessary. Article 1 of the EnWG takes up environmental concerns as an additional target for energy supply, but no further changes in the draft law stress the importance of environmental concerns. Concession treaties with exclusive clauses and demarcation treaties will not be allowed in the future.

A second draft of the EnWG review dropped unbundling and reduced third-party access allowance. In March 1994 the EnWG review was withdrawn by the BMWi, mainly because the municipalities declared open resistance.[12] Further, the majority of SPD-governed Länder in the Bundesrat signalled, that this EnWG review would be halted. The plan to review the EnWG during the 12th legislation period was dropped. The Minister of Economy said that German national legislation in the future should be developed parallel to planned Directives at the European level.

In February 1996, the BMWi presented in Parliament the next draft of the EnWG review. Its focus is again increased competition, by third-party access or new transmission lines. Demarcation treaties and exclusive clauses in concession treaties still will be abolished. The association

---

[12] Press release of Bundesverband der deutschen Gas- und Wasserwirtschaft & Verband kommunaler Unternehmen, March 2, 1994 in Bonn.

of German electricity stations (VDEW) reacted relatively critically. Together with representatives of national grid companies such as RWE and VEAG, the necessity for long transformation periods were stressed. The association of regional utilities (ARE) broadly agreed with the draft. Opposition came from the VKU, who said that the draft turned upside-down the framework conditions of energy law in Germany. The Association of German Towns and Communities pledged the continuation of closed supply areas by introducing a competition system at this level.

V. Prospective Reflections—Where is the System Going?

Critical reports like the Deregulation Commission's reports or the reports by the Enquete Commission on Climate Protection are not necessarily adopted and implemented by the Federal Government. Only some ideas from the many suggestions were taken up in the energy policy statement of the Federal Minister of the Economy (Bundesministerium für Wirtschaft, 1992), mainly in connection with EC reform attempts. Large utilities are opposing unbundling and other suggested structural reform attempts.

However, the large utilities have realised that the decades of stable monopoly in Germany are over. Therefore, slight reform steps are no longer averted, but are used as a chance to conquer electricity customers in other European countries. Regulated competition in electricity is seen as a lesser evil, and slowing down the development towards free competition will even make it possible to realise much more environmental protection in heat and power production. This opens a way for small-scale solutions such as decentralised CHP plants and demand-side management programs. Stagnating electricity consumption and increasing environmental consciousness within the ESI tend to alter the investment strategies of the big utilities. Big power station blocks will not be the solution in a situation of uncertainty for traditional supply markets.

The future of nuclear power in Germany is of decisive importance. In case of a red–green Federal Government (which may be the result of the next elections to the Bundestag) the phasing out of nuclear power in Germany will be on the agenda. This indicates the middle-range substitution of more than 20,000 MW base-load capacities.

Literature

Arbeitsgruppe Energie und Umwelt (1993) Auswirkungen des Elektrizitätsbinnenmarktes auf die Umwelt. Bericht für die 40. Umweltministerkonferenz, *Umwelt*, No. 9.

ARE (1993) Regionale Energieversorgung 1994–1995, *Tätigkeitsbericht der Arbeits gemein-schaft regionaler Energieversorgungs-Unternehmen*, ARE eV, Hannover.

Bundesministerium für Umwelt, Naturschutz und Reaktorsicherheit (1992) *Zur Novellierung des Energiewirtschaftsgesetzes—Defizitanalyse und Reformkonzeption aus umweltpolitischer Sicht*, Bonn.

Bundesministerium für Wirtschaft (1992) *Energiepolitik für das vereinte Deutschland*, Bonn.

Bundesministerium für Wirtschaft (1995) *Energie Daten 95. Nationale und internationale Entwicklung*, Bonn.

Bundesministerium für Wirtschaft (1996) *Die Elektrizitätswirtschaft in der Bundesrepublik Deutschland 1994*, Frankfurt.

Deregulierungskommission (1993) *Marktöffnung und Wettbewerb. Berichte 1990 und 1991*, Schäffer-Poeschel, Stuttgart, pp. 66–90.

Deutsche Verbundgesellschaft (1995) *Bericht 1994*, Heidelberg.

Hennicke, Peter *et al.* (1985) *Die Energiewende ist möglich*, Fischer, Frankfurt.

Jänicke, Martin (1990) *State Failure. The Impotence of Politics in Industrial Society*, Polity Press, Cambridge.

Jänicke, Martin, *et al.* (1989) *Ziele und Möglichkeiten einer stromspezifischen Energiesparpolitik in Berlin (West)*, Berlin.

Kristof, Kora (1992) *Dezentralisierung in der Elektrizitätswirtschaft*, Campus, Frankfurt/New York.

Mez, Lutz, Jänicke, Martin and Pöschk, Jürgen (1991) *Die Energiesituation in der vormaligen DDR. Darstellung, Kritik und Perspektiven der Elektrizitätsversorgung*, Edition sigma, Berlin.

Mez, Lutz *et al.* (1994) *Energiewirtschaft—Energiepolitik, Skript für den Weiterbildungs-studiengang 'Umweltschutz' der Humboldt-Universität zu Berlin*, 2. Aktualisierte Auflage, Berlin.

Monopolkommission (1976) *Mehr Wettbewerb ist möglich. Haugtgutocrten 1973/1975*, Nomos, Baden-Baden.

Das Parlament (1993) No. 32 (Schwerpunktthema Energiepolitik).

Schiffer, Hans-Wilhelm (1995) *Energiemarkt Bundesrepublik Deutschland, 5. völlig neu bear-beitete Auflage*, Verlag TÜV Rheinland, Köln.

Schmitt, Dieter and Heck, Heinz (eds) (1996) *Handbuch Energie*, Neske, Pfullingen.

Stelte, Michael (1994) Energy Data Bank, Berlin.

Vereinigung Deutscher Elektrizitätswerke (1995) *VDEW-Jahresbericht 1994*, Frankfurt.

Vereinigung Deutscher Elektrizitätswerke (1995) *Die öffentliche Elektrizitätsversorgung 1994*, Frankfurt.

Vereinigung Deutscher Elektrizitätswerke (ed.) (1995) *VDEW-Statistik 1994. Teil I: Leistung und Arbeit*, VWEW Verlag, Frankfurt.

Verband der Industriellen Energie- und Kraftwirtschaft Frankfurt (1995) *Tätigkeitsbericht 1994/95*, Essen.

Verband der Industriellen Energie- und Kraftwirtschaft (ed.) (1996) *Statistik der Energiewirtschaft 1994/95*, Essen.

Verband kommunaler Unternehmen (1995) *Kommunale Versorgungswirtschaft 1994/95*, Köln.

Walz, Rainer (1994) *Die Elektrizitätswirtschaft in den USA und der BRD. Vergleich unter Berücksichtigung der Kraftwärme-Kopplung und der rationellen Elektrizitätsnutzung*, Physika, Heidelberg.

Zängl, Wolfgang (1989) *Deutschlands Strom. Die Politik der Elektrifizierung von 1866 bis heute*, Campus, Frankfurt/New York.

Part Five

Comparative Analysis

Chapter IX
Electricity Policy Within the European Union: One Step Forward, Two Steps Back

ATLE MIDTTUN

I. Introduction

In his book *One Step Forward, Two Steps Back,* Lenin (1904), in a polemic against a more opportunistic minority wing of the Communist Party, speaks of revolution as a challenge of endurance, where it is necessary to be prepared to make extensive retreats before advancing again. Such endurance is also obviously needed in the forging of a common EU energy policy. Over the last decade, we have seen ambitious attempts by the European Commission to advance common liberal electricity and gas regimes, only to meet strong industrial and national opposition. In the next round, this has led to clear policy retreats and procedural entanglement and a complex set of negotiations, pointing towards outcomes which could be called a 'two steps back' solution, although this may exaggerate the retreat.

This chapter gives an exposé of some of the main lines in this development, and seeks to relate them to structural and procedural characteristics of the complex bargaining arena which constitutes the European Union. An integration of European electricity and gas markets challenges strong and entrenched industrial interests and deeply rooted national styles of industrial organisation. These two sectors are, therefore, probably among the most difficult to encompass within a common European regulatory regime. As the electricity sector was chosen to be a prototype for gas regulation, this chapter is primarily about electricity. However, the policy conclusions are largely also valid for gas.

The chapter has nine main sections. Section II spells out the analytical framework, conceptualising EU energy policy as a complex multi-level and multi-institutional bargaining game. Sections III–VII discuss various phases of EU electricity policy development, starting out with the post-war national infrastructure model (Section III), moving on to the mild voluntary co-ordination of strategies to meet the 1973/74 oil crisis (Section IV), followed by the lack of European strategies to handle the nuclear crisis in the 1980s (Section V), subsequently discussing the development of a strong European liberalisation policy (Section VI), and finally the retreat to a weaker bilateralist position (Section VII). Section VIII highlights the cognitive dimension of the liberalisation policy debate, and Section IX speculates about some of the determinants of future European energy policy.

II. EU Electricity Policy as at Complex Multi-level and Multi-institutional Bargaining Game.

The struggle to develop a common EU electricity policy can be conceptualised as a complex multi-level bargaining game across multiple institutional contexts. It is multi-level because decisions over regulatory principles are made at both national and EU levels. It is multi-institutional because decision-making may take place within legal, political, administrative and corporate sectoral arenas. It is complex and bargained because of the lack of clarity about the relative decision-making power of each of the levels and institutional arenas. The complex bargaining is clearly illustrated by the constant tension between national compromise in the Council of Ministers and the more visionary policy-making in EU-level policy organs like the European Commission and the European Parliament.

A core element in the bargaining around a common European energy policy can be formulated in terms of a simple 'prisoner's dilemma' game over the realisation of common European interest, given strong national stake-holders with different resources. One side of this game are the potential economic gains from pursuing collective integrated strategies. The other side are the extensive institutional and competitive differences between the member countries and their electricity industries, leading some of them to forsake European collective gains by sticking to closed, national strategies which prevent them from competitive international exposure.

The welfare potential to be reaped from an integrated European electricity market is considerable, and relates to potential gains from optimising systems at a higher level in terms of:

1. the economics of new combinations of production resources, for example of combining hydropower and coal-based electricity production;
2. the economics of load management across time zones;
3. the security of supply increases from multiple connections and multiple fuel systems, for example to diminish the problem of fall-outs of major nuclear stations, or the lack of hydro-power in dry years; and
4. the economics of optimal investment with a larger set of investment options.

The electricity sector is one of the largest industrial complexes in the modern economy, and extensive efficiency gains here would have a substantal impact on national economies.

The distributive implications of a competitively integrated European electricity supply are, however, just as striking. With industrial electricity prices ranging from 0.35 $/kwh in Norway to 0.93 $/kwh in Germany, there would obviously be heavy losers from competitive exposure. Looking at the countries involved, it is hardly likely that these price differences might be accounted for by differences in transmission, which would still remain under a competitive regime. Cutting down German electricity prices, which are presently around 56% and 57% above household and industrial EU averages (IEA...), respectively, to a median level would imply sizeable cut-backs with repercussions far into the public sector. Not only is the German electricity industry earning good profits, but municipalities are harvesting abundant concession fees and the coal industry is extracting extensive coal subsidies. Likewise, a scaling up of Norwegian electricity prices, which are presently around 46% and 54% below household and industrial EU averages, respectively, would provoke extensive consumer protest from both households and industry.

The two extreme positions of the European electricity policy game are thus on the one hand a collectively integrated regime with large potential benefits for the Union as a whole, but with severe consequences for established electricity industry in certain countries, and on the other hand a minimum common multiplum regime, which more or less maintains *status quo*.

The internal tension within individual countries between sectoral and national decision-making makes for a less clear-cut distinction between collective European and national interests than the simple 'prisoner's dilemma' model seems to imply. The introduction of a multi-level (European/national/sectoral, etc.), rather than a two-level (European/national) model implies the possibility of more complex alliance structures. One could, for instance, conceive of alliances forged by European-level decision-makers and 'sympathetic' national interests in a strategy to overcome the dilemma of collective action which Europe faces in this field. Consumer interests at the national level in countries with high electricity prices may in fact side with EU-level integration strategies, in spite of strong opposition from producer interests and those public interests which benefit from their high profits. The national position towards common European policy initiatives would in this perspective depend on the relative strength of the two pressure groups in national policy-making.

Furthermore, institutional differentiation inside the policy-making apparatus, within both the national and European policy levels, implies that policy-making in one institutional arena might be challenged by policy initiatives in another, adding new potential alliance structures and conflict dimensions to the overall policy game. The tension between the administrative and the political arenas at the EU level is a case in point. The European Commission has, on several occasions, proclaimed policies which the Council of Ministers has later rejected. On the other hand, the Commission frequently bargains with recursion to the legal arena as a fairly explicit threat. Similar tensions can be found at the national level. German competition authorities have, for instance, on several occasions sought to open up the institutional monopolies of the large electricity companies, but have been met by effective opposition at the political level.

III. Public Infrastructure Provision Within Closed National Systems

The dominant post-World War II model for electricity-sector organisation was characterised by the conceptualisation of electricity as a public infrastructure and part of the process of nation-building. It was a basic element in industrial policy, and an important service to be made accessible to all consumers. In this period, large investments were therefore made in electricity systems, and subsidies were often made available to expand electrification to remote rural areas. Theoretically, the public ownership model was legitimised by the concept of natural monopoly (Samuelson, 1954, Musgrave and Peacock, 1967) based on scale advantages and welfare arguments which have been discussed in chapter II of this book.

The conceptualisation of electricity as a public infrastructure with natural monopoly characteristics led to an organisation of the sector into publicly-owned institutional monopolies, dominantly operating as closed national systems. Co-ordination between these systems was undertaken on a voluntary basis, organised by sector associations like UCPTE and NORDEL. Trade between them was largely a matter of marginal exchange of surplus to balance nationally independent production systems, and exchange prices were usually based on short-term marginal costs. Depending on the resource base and national institutional traditions, some countries centralised the electricity system at the national level, whereas others anchored the electricity system organisationally at the regional and local levels. In all cases, however, the mandated public organisation had exclusive monopoly rights to supply customers located within its domain.

This institutional set-up helped develop highly influential sectoral organisation with close ties between regulatory and operative organisations with considerable political legitimacy and influence, allowing the industry large scope for controlling policy-making at the sectoral arena (Midttun, 1987). The ability to control its own policy environment, in combination with an expansionist orientation built in to fulfil the demands of industrial growth in the early post-war decades, led the electricity industry to over-investment in electricity production when demand growth subsided throughout the 1970s and 1980s. The fact that the balancing of supply and demand took place according to sector-internal forecasts without proper concern for price variables, and with non-commercial allocation of public funds, helps to explain this development (Baumgartner and Midttun, 1987).

The French nuclear programme is an obvious example. Here, government funding was made available for a huge investment in French over-production, which has made France a major net exporter of electricity to the rest of Europe. Another example is the ambitious Norwegian hydro-power programme, which has given Norway a leading export role in the Nordic arena.

The post-war model of electricity regulation left little role for European economic co-operation except for the promotion of nuclear power. Nuclear power had been a central part of EEC energy policy right from the start. In fact, the European Atomic Energy Community (Euratom) was signed along with the treaty of Rome in 1957. The aim was to develop technology to facilitate the growth of available resources in member countries. The goal for Euratom, according to Article 1 of the Euratom treaty, was thus:

... to contribute to increase the standard of living in member states and to the development of relations with other countries by creating necessary conditions for a quick creation and development of nuclear industry (Quoted from Nutek, 1991, p. 8).

However, because of the dominance of US nuclear technology in the early phases of European nuclear programmes, Euratom never succeeded in forging a common European policy for nuclear power.

The natural monopoly of the electricity system remained uncontested throughout the whole period, and remained a solid foundation for industrial regulation. In spite of the French and the Norwegian examples, the traditional regimes with their national focus led to an underdeveloped international trade. International regulation was based on a minimum common multiplum, involving such issues as technical standards and regimes for marginal exchange. The latter were generally negotiated on a bilateral basis. The low ambition for international regulation, and the fact that the participating nations and companies were not competitors but co-operators on the basis of mutual benefit, implied that bargaining problems were minimal. Decision-making in the international forums involved was generally consensus-based.

IV. The Oil Crisis and Mild Voluntary Co-ordination

The successful 'cartelisation' of oil producers and the ensuing price rise in 1973/74 triggered a demand for greater co-ordination of energy policies, particularly to diminish the dependence on oil and even natural gas. Measures to this effect were taken nationally, but also in collaboration within the European Community and the IEA (International Energy Agency), formed to meet these needs in 1974 after initiatives from the USA. This situation gave the EEC an opportunity to take a new co-ordinating role, but it was a mild co-ordination, within the framework of the traditional model. The public service orientation, the institutional monopoly and the basic closed-system character of national electricity supply remained unaffected. The diminished dependence on non-European fuel sources in fact strengthened the closure of the national systems on the input side.

In line with the planned-economy tradition which prevailed in the electricity sector, the European energy policy goals were specific and quantitative. The first goals were formulated in 1974 (Council Resolution of 17 September, 1974 concerning the new energy policy strategy for the Community OJ C 153, 9.7.1975), and then specified further in 1980. Common efforts within the European Community were also embedded in two Directives which limited the development of electricity production based on petroleum products and natural gas (directives from the Council Nos. 75/405/EEC and 75/404/EEC). The former Directive is still valid.

The fuel fuel-shift policy was in fact quite successful. Although the ambitious goals were not completely met, import from countries outside the EEC was halved during the decade between 1973 and 1983, and the energy intensity of industrial production fell by 20%. However, a later fall in oil prices delayed the implementation of these goals.

The fuel-shift and energy-saving policies have remained on the European Community policy agenda ever since the oil crisis. The energy policy goals of the EEC were codified in a Council Resolution (of 16 September, 1986 OJ C 241, 25.9.1986). The quantitative goals set forth in this document, to be fulfilled by 1995, included:

1. an efficiency increase in energy consumption of 20%;
2. a cut in oil consumption to 40% of total energy consumption;
3. an increase in the use of solid fuels;
4. a more competitive production of solid fuel;
5. oil and gas should constitute less than 15% of total electricity production;
6. production of renewable energy sources shall increase; and
7. the important role of natural gas in the energy balance shall be maintained.

The apparent success of this policy is probably due to a number of factors, including the broad consensus between national and European decision-makers over the goals, a choice of planned-economy means which suited the closed, national public-service model, in conjunction with price incentives which worked the same way. In this sense, the European fuel-shift and energy-saving policies were an example of the successful application of the subsidiarity principle to further common welfare.

It is important to point out, however, that one of the material premises was that there were no losers in this game within the European Community. The powerful electricity industry did not have to give up sovereignty, and nor did the national states. The European Community here acted as a facilitator alongside the IEA to help co-ordinate national sectoral efforts to reach national and sectoral goals.

V. The Nuclear Crisis and the Lack of a European Policy

In spite of the fact that the development of nuclear energy was defined as a central part of the European Community's energy policy, the nuclear crisis in the 1980s and early 1990s was definitely handled at the national level, with little or no European co-ordination. This stands in stark contrast to the successful international co-ordination of the previous fuel-shift policy. The intense politicisation of the nuclear issue obviously contributed to this outcome.

Throughout the 1970s and 1980s several west European countries built up nuclear programmes and nuclear energy was partly, within the Euroatom frame, seen as a major solution to energy crisis and energy scarcity. The programme gradually met forceful political protests, which in most west European countries led to stagnation, and in certain cases also to a scaling down of nuclear energy. Commercial exposure in Great Britain in itself also led to an abandonment of nuclear power, which now, in a commercially exposed energy supply system, has to be given extensive subsidies.

In most states, the nuclear conflict intensified at the end of the 1970s and the start of the 1980s, when major encounters between the nuclear-oriented establishment and nuclear opposition developed along both institutional and confrontational lines. Referenda were, for instance, held in Sweden and Austria, while violent clashes between protesters and police occurred in Germany and France. Nuclear protest was also partly linked to a broader environmental opposition, but this pattern varied from country to country

Political confrontations, with the most massive police intervention since the World War II in several European countries, were clearly too controversial to be delegated to the weakly politicised European Community institutions and to escape heavy engagement in the national political arena. The dominantly technical/administrative competence of sectoral decision-making arenas were obviously too restricted to solve this larger legitimacy crisis. Except in France, initial attempts to define the nuclear opposition as irrational and to reinstate the traditional values and evaluatory criteria of the sector institutions generally failed (Flam, 1984).

Except for France, the west European nuclear countries all ended up with a more or less *de facto* nuclear moratorium. Some countries, like West Germany and Great Britain, increased their nuclear capacity somewhat, but then ended up with a *de facto* moratorium. Sweden completed an ambitious nuclear power programme, even though it decided to phase nuclear power out in the future. Other states remained with a low or non-existent nuclear power capacity, because they had either cancelled their engagement in the nuclear business at an early stage (Norway), had withdrawn after an earlier commitment (Austria and Italy), or had frozen it at a low level but had maintained a nuclear option (The Netherlands). The diversity of solutions chosen and the intensity of national political debate indicated that this issue, at the core of European energy policy, was too controversial for European co-ordination. The most successful international co-ordination of the nuclear legitimacy crisis in fact came from the nuclear opposition, operating largely outside established party-political arenas.

VI. Towards a Strong European Liberal Electricity Policy

In the course of the second half of the 1980s and the early 1990s, vigorous steps were taken towards a liberal and more integrated European electricity and gas market. The goal was to include these energy sectors under the original European Community vision of 'establishing a common market and progressively approximating the economic policies of Member States' (Treaty of Rome, Article 2). This new policy definitely broke with many of the traditional 'axioms' of the old organisation of the electricity industry. It suggested a move from closed, national systems to an open, European system, from institutional monopolies to competitive interaction, from disparate national to common European regulation.

The arguments for such a move are, of course, the classical arguments for free trade, and of increased allocative and dynamic efficiency. Politically, this move was seen as supplementary to the parallel process of establishing a general internal market. The inclusion of the grid-bound energy sectors was seen as extending and reinforcing the European market process and European institutions.

The formulation of a strong common, market-oriented EC policy on grid-bound energy was launched in 1988 with the publication of the Commission's Working Paper entitled 'The Internal Energy Market' (Com (88) 238 final of 2 May, 1988). This document followed the pattern which had been used with great success in the 1985 White Paper on the internal market, applying it to the energy sector specifically.

Strictly speaking, the concept of an internal energy-market follows directly from the general concept of the EC's internal market, and from the basic rules of freedom of trade codified in Articles 30–36 of the Treaty of Rome. The new internal energy-market initiative was therefore more a question of the implementation of rules in practice than of a change of rules. In principle, trade in energy has not been explicitly/ formally exempted from the competition rules embedded in the Treaty of Rome, but these rules have traditionally not been applied to grid-bound energy, due to strong national energy policies and the sectoral autonomy of privileged companies.

The policy outlined by the EC Commission was in many ways a strong and confident policy, reflecting the apparent success of the general 1985 White Paper on the internal market, and perhaps the belief that decision-making on this issue in the Council of Ministers would, according to Article 100A of the Treaty of Rome, only need a qualified majority. This would have saved the proposal from the bargaining game which arises with a demand for unanimous decision-making. As already mentioned, the unanimity rule gives most power to the most disinterested actor, who is thereby in a position to severely obstruct any further implementation.

The first step towards implementation of the Internal Energy Market was taken by the Commission's proposal of the following Directives in 1989:

1. price transparency for gas and electricity (COM (89) 332);
2. transit rights for gas and electricity (COM (89) 334 and 336); and
3. co-operation on infrastructure and co-ordination of investments in energy projects of common interest (COM (89) 335).

These Directives were subsequently adopted by the Council of Ministers, but only after strong resistance from the energy industry and most Member States. After protracted negotiations the price transparency was accepted, though in a modified form (OJ no. L 185 of 17 July, 1990). The proposals on the transit of gas and electricity were approved by a majority vote, after long and tough negotiations, with several Member States still opposing the Directive. The final results of these Directives were limited in scope, but nevertheless seen by the Commission as an important step toward creating an internal energy market.

Following the three Directives, the Commission, on 23 October, 1991, in the optimistic spirit of a strong policy, adopted a set of guidelines for completion of the Internal Market in gas and electricity. The guidelines spelled out three stages: (1) implementation in 1991 and early 1992 of the three Directives mentioned above, (2) progressive elimination from 1 January, 1993 onwards of exclusive rights, and (3) the completion of the internal gas and electricity markets from 1 January, 1996 onwards, in the light of experience gained during stages (1) and (2).

The Commission continued the pursuit of its strong policy by tabling two proposals for Directives on 22 January, 1992, designed to abolish exclusive rights for the production of electricity and the construction of power lines and gas conduits. The two proposed directives also called for separation of management and accounting with respect to production and distribution activities in vertically integrated companies and, in addition, for gas and electricity transport and distribution companies to allow certain distributors and large consumers access to their networks—so-called 'third-party access' (TPA).

These two directives marked the culmination of the strong policy on the internal energy market, ironically as the general internal market was being put into practice. In November 1992 the Council decided that it could not accept the measures as presented, especially TPA, and asked the Commission to modify these proposals. Thus, stage (2) was not able to begin on January 1, 1993 as envisaged by the Commission.

As part of the strong policy, the Commission also took initiatives to apply EC monopoly rules to agreements between energy companies. In 1991 the Commission thus intervened against an agreement between SEP, the company responsible for co-ordinating electricity production in The Netherlands, and the distribution companies which granted SEP the exclusive right to import electricity. This was the first time that the Commission reacted to limitations of electricity trade with reference to the competition rules in the Treaty of Rome, more precisely Article 85 (Nutek, 1991).

In another case in the beginning of the year, it was made clear, by a verdict in the EC court that the Commission may make use of Article 90 in order to abolish import and export monopolies for telecommunication supplies (Nutek, 1991). With the support of this verdict, the Commission has requested member countries to account for the national laws which motivate export and import monopolies for electricity and natural gas. The response to this request was, however, reluctant, and opposition from the electricity and gas industries to a quick dismantling of restrictions to trade has turned out to be prohibitive to change.

As part of the strong policy for an internal energy market, the Commission has, since 1988, also adopted a more stringent attitude to coal. Subsidisation of national coal production is to be transparent and part of restructuring and rationalisation programmes aimed at closing uncompetitive production. Only subsidisation which meets these criteria and which therefore is limited in time will be approved by the Commission. On this basis, the Commission has intervened in Germany's discussions on the future of the country's coal policy (Mez, 1994). Like its competition policy regarding the energy sectors directly, the Commission's coal policy was based on a legal foundation. The Paris treaty which constitutes the basis for the Coal and Steel Union prohibits the use of duties and other trade restrictions for coal trade between member countries, as well as subsidies and 'cartelisation' of the coal producers (Nutek, 1991).

The strong liberalisation policy was, as we have seen, vigorously pursued by the European Commission, playing an 'avant garde' role, and only more reluctantly by the European political institutions, the Council of Ministers and the Parliament. The dramatic break with the traditional closed, public-service model obviously led to less enthusiasm at the sectoral and national levels in countries like France, Belgium and

Denmark, where the adherence to a nationally focused planned-economy and public-service model still was strong.

VII. The Retreat to a Weaker Bilateralist Position

Several events during the early 1990s indicated that unlike the general internal market policy, the strong internal energy market policy was running up against massive opposition which it was unlikely to break through. As the Commission strove to implement the more ambitious elements of its strong energy market policy and to apply the competition rules embedded in the EC treaty, it became clear that decision-making was becoming blocked over the more controversial element of the internal energy market. Decision-making took more and more the character of negotiations under threat of veto from single actors, where the least interested parties (France, Spain and Italy) had most to say. The hope of reaching through with majority decisions or appeal to formal rules proved useless.

The rather blatant setting aside of Community principles of free trade and the unwillingness to override minorities by strong majority-based policies reflect the strong reservations which Member States have when it comes to sharing sovereignty over this field. The great diversity in resource endowments and industrial policy traditions imply that various national electricity and gas industries would face international competition with very different capabilities. Given their strong influence in national policy-making, and the reluctance of many national gas and electricity companies to give up their monopolies, the strong liberal European energy policy was up against powerful opposition.

At a more general level, the lack of trust *vis-à-vis* European integration, demonstrated in the referenda in Denmark and France in 1992 over the European Union, may also have inspired a reversal to a less controversial, weak, subsidiarity-oriented policy. In fact, deep disagreement on integration policy among Member States made it impossible to include energy as a common policy area in the Maastricht Treaty. The first draft included a chapter on energy policy, which was withdrawn in the final round of negotiations. Energy policy, it was felt, should remain an area where the principle of 'subsidiarity' applies. However, some elements of the original energy chapter survived in the Maastricht Treaty: Articles 129b and c in the chapter on trans-European networks state that the EU shall promote the development of energy infrastructure and the opening up of energy transport. EU structural funds were designed to be used for this purpose.

The first step towards a weakening of the internal energy-market policy was taken by the Commission in response to the Council's rejection of the two stage-2 Directives. Accordingly, the Commission handed to the Council on the 8 December, 1993 two modified proposals for directives. The essential differences from the 1992 proposals to be found in these new texts were:

1. *negotiated* instead of *full* third-party access to networks;
2. clear references to *public service* obligations;
3. greater harmonisation in order to ensure efficient functioning of the internal gas and electricity markets; and
4. a tender system instead of a system of licensing for new electricity production and transport capacity.

The controversial nature of the liberalisation of grid-bound energy gave the European Parliament increased influence over development, which in turn further weakened the liberal policy position. Some of the basic premises for the European Parliament's policy position, were (Svindland, 1995):

1. that it was more important to harmonise the national electricity systems than to introduce more competition between the European companies;
2. that harmonisation and deregulation were to be introduced step-wise;
3. that progress in the introduction of a common internal electricity market must rest on a previous harmonisation of energy policy; and
4. that the principle of subsidiarity must have priority in the harmonisation process.

The Committee for Industrial and Social Affairs delivered its report to the European Parliament on 27 January based on the above principles, which subsequently gained support in Parliament. On the 17 November, 1993 the European Parliament thus voted for a policy which emphasised that other policy areas, particularly environmental policies and taxation policies, should be harmonised before the national electricity markets were integrated. The parliamentary decision emphasised the right of Member States to oblige their electricity companies to a 'public service' attitude, and allowed actors vested with concessionary rights to protect and maintain monopolies. The Member States were left to choose between liberalisation through public tender or through transparent, non-discriminating rules for the treatment of applications for concessions. Third-party access was proposed under restricted conditions of negotiated agreements, and only for large industrial segments. TPA was not to be introduced for distribution companies with a centrally planned electricity model (Svindland, 1995).

In spite of the obvious difference between the proposals of the Commission and the Parliament, active attempts were made to harmonise them. The European Parliament is usually the weaker party in such negotiations, as it only has an advisory mandate. The Commission has therefore been able to reject Parliamentary decisions with reference to the fact that they do not satisfy the intentions of the Treaty of Rome, particularly when central parts of the Treaty are concerned. With the recent strengthening of the European Parliament as part of the general process of European integration, and given the strong national positions on electricity policy, the Commission was nevertheless willing to reconsider its proposal. The strong engagement from the Council of Ministers in this issue obviously also contributed to further the willingness of the Commission to compromise.

Following months of discussion, the Council laid down a number of general principles and reached an agreement on 29 November, 1994 on four points of the electricity sector dossier (Belaud, 1995):

1. Member States should have a choice between licensing procedures and/or a tender system when setting up new electricity production facilities;
2. there should be separate accounting for production, transport and distribution activities in vertically integrated companies, in order to avoid discrimination, cross-subsidies, and distortions of competition, controlled by independent bodies;
3. provisions in the Directive 'relating to network operators should be confined to what is necessary to ensure implementation of the internal electricity market in an unbureaucratic manner'; and
4. there should be acceptance of public service obligations—in those Member States which impose them—but on the conditions that they were clearly defined as transparent, non-discriminatory, and capable of being monitored, and that they also comply with the requirements of EU competition rules

After the French launching of the so-called '*single-buyer/supplier principle*' (hereafter abbreviated as the single-buyer model) to protect EDF's monopoly position, much of the European electricity regulation debate has concentrated on the harmonisation of this position and the Commission's concept of negotiated TPA. Initially, the Commission condemned the French single-buyer model as a further weakening of its proposal, calling it 'an import monopoly by another name'. Nevertheless, it later had to seek a compromise. This was obviously due to support for the French model among member countries. The single-buyer model has enjoyed support from France, Greece and Ireland. The Netherlands wants TPA, but wishes to keep planning intact.

Denmark stresses the need to uphold environmental considerations. German electricity companies have, however, been concerned with the distortion of competition which they fear would result from the French model when interacting with a modified third-party model.

TPA, the Commission's preferred method in its initial proposal, means that electricity producers would be able to sell directly to eligible customers—basically to industry—by negotiating access to the electricity grid, allowing such customers to choose their electricity supplier form anywhere across the European market. The single-buyer model provides for a single entity which would buy and sell all the electricity generated, under competitive conditions, and operate the network. All customers would then buy from this 'single buyer', who would manage all electricity deals, with producers and consumers negotiating between themselves on import arrangements only.

In spite of the Commission's reluctance, the Council of Ministers, at its meeting in November 1994 explicitly requested the Commission to examine and outline the anticipated consequences of how the Commission's proposal of negotiated TPA could exist alongside the French single-buyer idea, and to produce a common position to be reached as soon as possible in 1995.

When the Commission presented its working Paper in early 1995, it maintained its sceptical position. It stated that the two models 'are so different that they cannot provide an equivalent degree of market access, nor reciprocity between the systems.' The Commission sees the single-buyer, with its effective import monopoly, as running contrary to Article 30 of the Treaty of Rome. When later persuaded to accept the single-buyer model as parallel to the negotiated TPA model, the Commission has tended to insist on conditions which would undermine the single-buyer's monopoly position.

The EU Commission's White Paper on energy policy, approved by Brussels in December 1995 (Commission of the European Communities, 1995), takes a pragmatic attitude and avoids strong policy formulations. It sets out a five-year energy work programme which can be implemented under existing Treaty instruments.

As far as the Intergovernmental Conference is concerned, considerable ambiguity remains over the issue of whether or not the rewriting of the Maastricht Treaty should include a new chapter on energy. The European Parliament seems strongly in favour of such a chapter, which is illustrated by its declaration on 13 March, 1996 that:

In order to achieve sustainable development, it is essential to establish the competence of the European Union in the field of energy by creating a new chapter on energy in the Treaty, where energy policy aspects of the European Coal and Steel Community and Euratom treaties and other energy policy considerations should be integrated within a common energy policy framework, helping to ensure overall co-operation with regard to security of supply and environmental protection within the internal market framework (Commission of the European Communities, 1996).

The enthusiasm for a special treaty on energy is, however, not shared by representatives of Member States. In its final report on the agenda for the Intergovernmental Conference, a majority in the Reflection Group has reiterated the view that it is not necessary to add an energy chapter to the Union Treaty (EC Energy Monthly, December, 1995). The Commission has also deemed it wise to leave this issue open in its Energy White Paper (Commission of the European Communities, 1995).

In the more specialised forum of the Council of Ministers, the continued debate over TPA *versus* the single-buyer principle has for a long time been running into a deadlock. The compromise suggested by the Spanish Presidency was to offer Member States the choice to either open up their supply markets by gradual and limited distributor access or to implement a similar opening for industrial and commercial consumers (EC Energy Monthly, November, 1995). In both cases it would be possible to tie up 70% of the new market by long-term contracts. This proposal has been rejected as being both too weak for the liberalisers and too strong for the public-service oriented members.

The French Government, after heavy strikes by public-sector workers, including those of Electricité de France, has found it difficult to compromise on electricity liberalisation. In Germany, the Bonn Government has been concerned about the lack of reciprocity or equal market access between the two countries if France operates the single-buyer model and Germany opts for negotiated third-party access. At the same time, the Government is under strong pressure from its industrial lobby to liberalise in order to achieve lower energy prices.

After a apparent loosening-up of the energy policy position of the Franco-German summit in Dijon in early June 1996, the Energy Council, in its meeting of 20 June made steps in a liberal direction, but by continuous and gradual means. The Council reached an agreement to open up 22% of the energy market to competitive trade by 1999–2000, and by 32% in 2003 (European Commission, 1996).

VIII. The Conceptual Dimension

The struggle over European electricity policy is not only a question of interests, but also over the cognitive models and the social construction of reality (Berger and Luckman, 1976). Besides the diversity of interests, there has therefore been a dispute over principles. Commercial liberalism has been confronted by the concept of public service, multilateral open models have been confronted with bilateral negotiated co-ordination, etc., reflecting the debate's deeper cognitive roots in national values and traditions.

Cognitive models and the social construction of reality are obviously often linked to underlying material interests. However, the conceptual debate also has a logic of its own. A clever conceptual formulation will further given interests by translating them into generalisable arguments within relevant legal, administrative and political settings. Furthermore, conceptual redefinitions may provide vehicles for renewed negotiations between seemingly contradictory interests, for instance by pointing out models for compromises which can reach broader collective acceptance.

By potentially providing ways around to the problem of collective action, the conceptualisation process may provide much of the dynamics of the bargaining process. The conceptual development from a competitive market with third-party access, over negotiated third party access, to the single-buyer concept (which we shall return to later) implies a sequential introduction of a European policy with milder and milder intrusion into national decision-making, moving the compromise further and further towards the minimum common multiplum point. However, how far from the original vision of a fully integrated common market these negotiations will end up depends in part on the success of devising conceptual tools.

The protracted conflict of interest over the liberalisation of electricity policy has clearly led to a parallel blooming of conceptual ingenuity. From a rather clear-cut liberal ideal type, the Commission has had to move towards a more blurred concept of 'negotiated' commercial solutions. The French launching of the single-buyer model helped give an intellectual/ideological framework to a policy position which was threatening to become reactionary. The confrontation over these two principles has again given rise to interesting reflection on integrating multiple types of regulatory models within a common trade context. In understanding the European Union debate, it is necessary to plunge a little deeper into this matter.

VIII.A. *The strong liberal model*

The rationale behind the strong EU liberal policy proposed in the late 1980s for competitive market organisation was, of course, the classical aim of marginalistic theory, to achieve optimal allocation of resources and thereby maximise welfare. As described in chapter II, the early public finance theory dealt with electricity *in toto* and, because of the natural monopoly character of the grid, advocated public monopoly. Later, more refined versions of regulation theory, however, advocated splitting the electricity system into two parts. On the one hand, there would be a publicly monopoly-regulated grid system, with full access for all market actors, and with supervised pricing. On the other hand, there would be a competitively regulated production, trade and distribution system, giving room for optimisation through marginalist adaptation.

The Commission's strong liberal model sought to apply modern regulation theory to the electricity sector's realities in a systematic manner, advocating on the one hand monopoly regulation of the grid with transparent pricing and third-party access, and on the other hand general competitive regulation of production, wholesale trade and distribution.

VIII.B. Negotiated TPA

The Commission's retreat to a weaker, negotiated third-party access position was obviously a fallback position to an analytically much more unclear regulatory model. On the generation side, the negotiated TPA implies that the Member State can choose between a system based on licensing or tendering for new plant. Under a tender system, industrial auto-producers must be allowed to construct new capacity, as they will be outside the central dispatching system. The same is true for independent power producers, although defining such producers is still the subject of debate.

Under both variants of the negotiated TPA model, both foreign and domestic generators must be able to negotiate access to the system to supply large industrial customers, defined as 'a final consumer buying electricity for his own use whose consumption exceeds 100 GWh or a lower quantity as may be specified by the Member State, or distributors 'on the basis of voluntary commercial agreements'. Electricity producers must be able to negotiate access to supply their own premises and subsidiaries in the same Member State or abroad, and foreign producers must have access to the grid if they win a tender contract (EC Monthly, February, 1975, ECE 72/1 and ECE 72/2).

VIII.C. The single-buyer model

The French single-buyer model is basically a defence of a planned-economy approach, with some formal openings for administered competition. The single-buyer will act as a planning and licensing authority: it will prepare tenders, specify load demand, required availability and the duration of contracts (in the case of import), and location. Choice of fuel and technology may also be specified, except for import.

In spite of its basic planned-economy orientation, the single-buyer model also seeks to incorporate certain aspects of liberal governance. As the transmission part of an integrated utility, the single-buyer may not give the production unit of the same company a favoured status in the tender procedures, such as access to information about competitors. The single-buyer model here assumes that an independent regulatory body, defined by individual Member States, will make the final judgement on the winner of tender contests. As far as the public grid is concerned, the single-buyer/supplier model fully advocates a planned-economy approach, and gives monopoly to the single-buyer/supplier.

On the supply side, the single-buyer/supplier model remains a planned monopoly, except for large industrial customers. Distributors are assumed to be forbidden by law to shop around for their electricity, either internally or *via* imports. Large industrial customers can either import power *via* a direct line with a foreign supplier, or negotiate with the single-buyer/supplier to import power *via* the public grid. In the latter case, the industrial customer must strike a deal with a foreign supplier, presumably undercutting prices offered by the single-buyer/supplier, and then sell it back to the single-buyer/supplier at a price between the two (EC Monthly, February, 1975, ECE 72/1 and ECE 72/2).

VIII.D. Mixed regulation

The existence of the negotiated-TPA and the French single-buyer/supplier model side-by-side has created a need for conceptualising mixed regulation, where different regimes can exist side-by-side with trade between them. Since the French launch of the single-buyer/supplier model, mixed regulation has been a central issue in the European regulatory debate.

The concern, under a subsidiarity-oriented regime, is how the two models are open to competition, and what effects competitive interaction between firms in two countries regulated by two different models would have on national electricity industries and on national trade balances.

A study by the Energiewirtshcafttliches Institut of Köln University in Germany for the Commission argues that, comparing the single-buyer model and the negotiated-TPA licensing option, there is more scope for supply competition in a negotiated-TPA system. However, the Institute suggests that there is more potential for competition in generation under the single-buyer system.

Some of the issues which will be crucial to the functioning of the two systems are, for negotiated TPA, the complexity of wheeling contracts and the problems of negotiating access to the grid, and for the French proposal, how to ensure strong barriers between the single-buyer as a planning and tender-specifying authority and its production interests (EC Monthly, February, 1995).

The conceptual level has a dynamic of its own. The blockage of the Commission's strong liberal position obviously necessitated a conceptual reformation to allow negotiations to proceed. The negotiated-TPA solution clearly has a potential in this adaptation toward a minimum common multiplum through the interpretation of the terms of negotiation. At the same time, negotiated TPA has the potential to develop into full TPA over time, as the Commission might gradually push for stricter terms of negotiation. This probably also motivated the strong French opposition to negotiated TPA.

The introduction of the single-buyer model as a conceptual tool has forced the other parties to relate intellectually to a more genuine, French, planned-economy position at a point when a simple veto to TPA would have seemed to be reactionary. This conceptual innovation has effectively delayed the liberalisation policy process, and may legitimise models compatible with domestic policy and industrial configurations of individual Member Countries.

As opposed to the negotiated-TPA model, the French single-buyer model would restrict liberal free trade in principle, and not only modify it temporarily in practice. An acceptance of this model would therefore be a major step towards legitimating planned economic governance within a liberal-market context. In order to make the single-buyer model more acceptable, French negotiators have recently launched the so-called 'positive reciprocity solution,' which implies that those countries which do not want to open up their distribution markets are not allowed access to the distribution markets of other Member States. However, EC lawyers are still worried that such a solution would be against the free movement of trade enshrined in the Treaty (EC Monthly, December, 1995).

VIII.E. Regulated gradualism

The June 1996 meeting of the Energy Council, marked a moderate breakthrough in the stalemate between the TPA and the public-service positions. By adapting a gradualistic rather than an 'in principle' position, the Council has obviously found it easier to compromise. The very cautious pace of market opening (up to 33% in ten years) also indicates a soft tone *vis-à-vis* nationally vested interests, which helps to make the compromise digestible even to public-service oriented members.

Furthermore the liberal market effects of the reform are limited by:

1. the exclusion of distribution companies from eligibility under the terms of the directive, meaning they are constrained from access to a competitive choice of supply;
2. the ability of Member States to choose between the negotiated third party access and the single buyer system for access to customers;
3. the possibility that integrated utilities can limit their unbundling to financial accounting measures, rather than physical or managerial separation;
4. the ability of Member States to limit competition in generation, allowing them to continue controlling the construction of new capacity and the fuel mix in generation, reducing the potential for new independent power producers to establish a presence in European markets.

IX. Energy and Beyond: Future Determinants of European Electricity Policy

From the rather messy state of compromises and mixed regulation, where and how will the European electricity market move? There is hardly any analytical basis for a definite answer. All we can venture is an overview of some of the major forces, interests and possibilities involved.

We have initially described European electricity regulation as a complex, multi-level bargaining process where actors at national and European levels seek solutions within different institutional arenas. The answer to the future development of the electricity regulation regime therefore lies both in the choice of arenas and in how this affects the bargaining power of the interests involved.

X. A The drive for integration

The overall development of European Union decision-making will, of course, also be a major institutional factor in the shaping of European energy policy, where a strengthening of the integration process will obviously strengthen the bargaining power of European Union actors *vis-à-vis* their national counterparts. The European integration process may also eventually lead to the development of more specified rules for electricity-sector decision-making, which may eventually change decision-making from a gaming process towards a more institutionally prescribed process.

There seems likely to be a continued institutional integration around economic governance, as we see a deepening of the Single European Market (SEM). As observed by Nugent (1994), this development does not only take place in response to the will of European countries to subsed decision-making to a supranational organisation like the EU. Much of the cause of decline in national powers is a response to an increased internationalisation of the economy, where the loss of sovereignty which arises from supranationalism is counter-balanced by the collective strength of political governance over market forces, gained by the EU as a whole. This process is most probably self-reinforcing, and is hence likely to create a considerable pressure for further economic integration, with implications also for the grid-bound energy sectors.

In addition comes the present development in the EU's Nordic fringe. With the new Member States Sweden and Finland joining Norway to form a liberal Nordic electricity market, the EU's liberal faction does not only contain the UK as a lonely outsider, but also a collaborative experiment by three countries to put into practice liberal principles which supersede the commission's strong ambitions.

Given the strong French resistance, there does not seem to be room for immediate dramatic policy change, at least as long as the utility–municipality coalition prevails in Germany against further liberal change. However, if the Nordic experiment turns out to be positive, this will obviously strengthen the Commission's bargaining position and serve to speed up the slow pace of liberalisation following the Energy Council's compromise in the spring of 1996, and subsequent decisions in the European Parliament.

IX.B. General competition rules

The deepening of the internal market, and the pressure from this process and the Nordic electricity-market experiment, will also increase the relevance of applying general competition rules to grid-bound energy. This may transfer the electricity industry to a broader decision-making arena, where the electricity industry has less policy-making control.

The Commission has already threatened fairly openly by referring the electricity liberalisation issue to the European Court under competition rules. First steps have already been taken in infringement procedures against six, and later (five) countries for maintaining export/ import monopolies. Later, on 27 April, 1994, the European Court of Justice, in its ruling in the Almelo *versus* Isselmij case (C393/92) clarified that the competition rules (Articles 85 and 86) apply to the electricity sector, and that electricity should be treated as a good, not a service. However, it did at the same time recognise that the full impact of the Treaty rules on competition can potentially be tempered where energy distribution companies are entrusted with public service duties. (Hancher, 1995).

In the same ruling, the Court also stated that Article 37, with requires Member States to ensure that state monopolies of a commercial nature do not operate to discriminate against imports and exports, can apply to situations where imports are restricted as a result of state-imposed measures (Hancher, 1995: EC Monthly, May, 1995). This has prepared the ground for the Commission's infringement actions against the five (Denmark has in the mean time removed Dangas's import monopoly).

IX.C. *Pressure from telecom and trans european networks*

To the forces which may influence the future development of European electricity and gas policy are pressures from other policy areas such as telecommunications, where structural characteristics allow for analogical transposition of regulatory solutions to grid-bound energy. As pointed out by Hancher (1995; EC Monthly, August, 1995) the overall progress of liberalising the telecommunications industry has been far more speedy. Under legislation now being devised by the Commission, all telephone services available to the domestic consumer will be opened up to competition by 1998.

A key to the success of liberalising the European telecommunications industry lies in the advance of technology, which has allowed telecommunications services to be supplied independently from the operation of the networks. This has made it easier for the Commission to take a step-by-step approach to opening up the sector. By separating out certain telecom markets and subjecting them to gradual liberalisation without having to confront head-on the problem of public service in its totality, the Commission has been able to reach European consensus for a liberal policy development. Opening up these 'value-added' areas such as fax and data transmission has thus acted as a lever to prise open the rest of the monopoly system (Hancher, 1995).

With this strategy, the Commission is reaching towards a fully fledged liberal reform, much along the lines of the 'strong' liberal policy position spelled out in the energy-policy field. A second stage of liberalisation, opening up basic telephone services and introducing competition to the entire network infrastructure is now conceptualised, and plans are also discussed for a Brussels-based European telecommunications agency.

Both on the general policy level and in practice, the emerging regulation of the telecommunications industry may set standards which may also come to influence the grid-bound energy industries. As liberalising the telecommunication sector in practice now seems to precede the more slowly evolving process in energy, principles developed in regulating telecommunications may provide important precedents for how such issues could be tackled in the future for energy. As Hancher (1995) notes, from a legal perspective, there would be little difficulty in extending the concepts upon which telecom-market liberalisation has been constructed to the energy sector. She argues, for instance, that the experience in the telecom sector lends extra weight to the Commission's threat of using Article 90(3) to prise open the power sector if progress is not reached soon on the internal energy market proposals. Political reluctance in the energy field may therefore imply that the rules for European infrastructure regulation will be developed in another sectoral arena.

The adoption of a gradualistic market-opening strategy in June 1996 by the Energy Council and subsequent decisions in the European Parliament must be seen against this background, as well as against one of the fairly open threat of active use of competition rules.

References

Baumgartner, T. and Midtunn, A. (eds) (1987) *The Politics of Energy Forecasting*, Oxford: Clarendon Press.

Belaud, J. F. (1995) *Energy in Europe*. Brussels: club de Bruxelles.

Berger, P. L. and Luckman, T. (1976) *The Social Construction of Reality*, Middlesex: Penguin Publishers.

Commission of the European Communities (1995) *An Energy Policy for the European Union*, Document COM (95) 682, 13.12.95, Brussels.

Commission of the European Communities (1996) *Parliament Resolution, Parliament Plenary Session, 13 March*, Brussels.

EC Energy Monthly (1993) *Distributor access dogs run-up to Energy Council*, 20 November, 1995, Issue 83, p. 83/1.

European Commision (1996) *Results of the extraordinary meeting of the Energy Council, Luxemburg, 20 June, 1996*, BIO/96/305, 21 June, 1996, Brussels.

Flam, H. (1994) (ed) *States and Anti-Nuclear Movements*, Edinburgh: Edinburgh University Press.

Hancher, L. (1995) 'Viewpoint: The hare and the tortoise. Comparing telecoms and energy liberalisation'. *Financial Times EC Energy Monthly*, August.

IEA (1993) *Electricity Information—1992*, OECD/IEA 1993, Paris, OECD.

IEA (1996) *Energy Statistics of OECD Countries (1960–1994)*, OECD/IEA, Paris, OECD.

Lenin, V I. (1904) *One Step Forward, Two Steps Back*, in Lenin Selected Works, Vol. 1, Moscow: Progress Publishers.

Mez, Lutz et al. (1994) *Energiewirtschaft–Energiepolitik, Skript für den Weiterbildungsstudiengang 'Umweltschutz' der Humboldt-Universität zu Berlin*, 2, aktualisierte Auflage, Berlin.

Midtunn, A. (1987) *Segmentering, institusjonelt etterslep og industriell omstilling: Norsk kraftutbyggings politiske økonomi gjennom 1970- og 1980-årene*. Avhandling for Filosofe Doktosexamen ved Universitetei i Uppsala.

Musgrave, R. A. and Peacock, A. T. (eds) (1967) *Classics in the Theory of Public Finance*, New York: St Martins Press.

Nugent, N. (1994) *The Government and Politics of the European Union*, London: Macmillan.

NUTEK (Swedish National Board of Industrial and Technical Development) (1991) 'EG och energin'. Stockholm: Allmanna forlaget.

Samuelson, P. A. (1954) *The Pure Theory of Public Expenditure*, The Review of Economics and Statistics. Vol. 40, pp. 387–389.

Svindland, E. (1995) *Tysk elektrisitietsproduksjon, forbruk og energipolitikk: Et grunnlag for norsk elektrisitetseksport?* Delrapport til Energiforsyningens Fellesorganisasjon, Gassgruppen.

Chapter X
Regulation Paradigms and Regulation Practice: a Comparative Review

ATLE MIDTTUN

I. Introduction

North-western Europe has seen a major shift towards more liberal electricity markets in the first half of the 1990s. The English and Welsh privatisation in 1990, followed by the Norwegian market reform in 1991, put into early practice principles which were also embedded in the European Commission's plans for an internal energy market. The European outlook of 1992 was therefore in many quarters one of liberalistic optimism, and a number of nations, including Sweden and Finland, were preparing to marketise or at least soften monopolistic control such as in The Netherlands.

However, there were strong exceptions. France, under the confident guidance of Electricité de France was, by the mid 1990s, still determined to pursue a centralised monopolistic policy, and the German electricity industry remained capable of resisting reform attempts due to its strong alliance with municipal authorities. Denmark was also firmly determined to maintain its planned-economy approach throughout the first half of the 1990s, although with some flexibility to accommodate local windmills and combined heat and power production. However, challenged by the advent of the Nordic internal market following liberalisation in Finland and Sweden, Denmark is also in the process of developing some kind of wholesale free trade.

Observations from the previous case studies raise two intriguing questions: why did the liberal shift of the early 1990s take place, after a long period of relative institutional stability? On the other hand, why was the liberal shift so selective, and why can we observe such extensive differences among countries?

There are many possible answers to these two questions. One answer to the first question obviously relates to the development of regulation theory. As pointed out in chapter II, we have seen extensive theoretical refinement, and the regulation professions have left the early post-war dichotomy of market or hierarchy for a far more refined and extensive regulatory menu. The new regulation theory has allowed us both to 'dissect' previously bundled services and to expose parts of them to regular market competition, as well as to regulate the remaining parts with a more refined regime of specially designed regulatory tools.

Another possible answer lies in the internationalisation of the economy and in the process of European integration. The pressure for competitiveness, both within and outside Europe, has put pressure on national electricity industries to supply power to competitively exposed electricity-consuming industries on advantageous terms. The process of European integration has also provided a promising arena for international trade in electricity for electricity producers with competitive production costs. Further integration of the European electricity industries through extended inter-European trade may also have appeared desirable to national electricity industries after the failure of several national systems to develop nuclear mono-cultures when this option became politically and economically unattractive.

Part of the answer to the question of why liberal market solutions became so popular is, of course, also politically related, and relates to an upsurge in neo-liberal ideology, with its most prominent advocates in Reagan in the United States and Thatcher in Great Britain. The political dimension of electricity regulation is, however, dealt with in the next chapter and will not be elaborated upon here.

As far as the second question presented above is concerned, we must again revert to our previous discussion on path dependency. As already mentioned in chapter II, a central element in the path-dependency perspective is that industrial systems cannot develop independently of previous events (David, 1993). Local positive loops serve to propagate traditional patterns into future strategic decisions. This implies a development with several equilibrium points, where small events at one point in time may play an important role for future development by determining the course of a long-term development.

Path dependency may here relate both to policy traditions, institutional and industrial structures, and to combinations of the two. The strong political commitment to public service and national political control in the French culture, combined with the monolithic organisation of the French electricity supply through EDF, provide a good example of a political–institutional configuration with apparent internal stability and considerable inertia to liberal change. The massive resources controlled by the EDF, Europe's largest electricity company, also gives the French position a central role in European development.

The far more decentralised, municipal ownership of the Norwegian industry, on the other hand, exemplifies a case that has lent itself more easily to market competition, once the pragmatic political decision was taken to move the system into a more market-oriented mode. As described in chapter IV, the tension between centralistic state ambitions and municipal autonomy in Norway illustrates how an element of institutional instability, given pragmatic liberal political reorientation, provided an opening for a new liberal regime.

In addition to differences in path dependency, the willingness to liberalise may also to some extent relate to assumed differences in competitive power, and expected gains from a wider European liberal market integration. In so far as electricity prices (exclusive of excise tax) reflect average production costs, the Norwegian, Swedish and Finnish electricity industries would clearly be in a better position than those in Germany, France and Britain (Fig. X.1). For hydro-based systems with a large reservoir capacity, such as Norway, a larger international liberal market system would also provide profitable income from load management.

This chapter summarises the patterns of European electricity regulation under the three main groups focused upon in this book: the

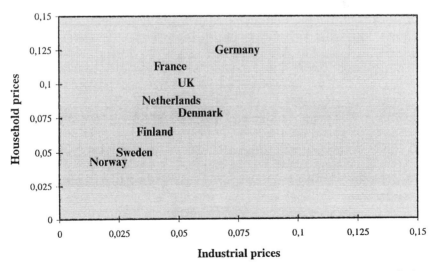

Fig. X.1. Electricity prices (1994) in ECU/kWh (excluding excise tax). *Source:* Appendix 2.

liberalisers, including Great Britain, Norway and later Finland and Sweden; monopolistic systems, including the French public-service oriented system and the 'cartelised' system in Germany; and systems in negotiated transition such as in The Netherlands and Denmark. The European Union's various positions over time are discussed within the main groups to which they respectively belong.

Both questions: why the liberal shift of the early 1990s took place, after a long period of relative institutional stability, and why the liberal shift was so selective, leading to such extensive differences among countries, are answered with reference to each of the main groups and countries within them. The last Section of the chapter draws some general implications for regulation theory.

II. The Drive for Liberal Regimes

As described in chapters III and IX, the drive for electricity liberalisation in Europe surfaced in the British Conservative Party's manifesto in 1987 and in the European Commission's Working Paper 'The Internal Energy Market' a year later. The British Conservative Party's initiative was a first step towards a dramatic privatisation policy which was put into practice in April 1990.

The British privatisation initiative was soon followed by a Norwegian market reform which was approved by parliament in June 1990 and implemented in January the next year. By early 1991 Europe therefore had two operative liberalised systems, and a strong liberal policy commitment by the EU Commission, and many observers believed that western Europe was facing a landslide in electricity regulation towards a liberal regime.

As opposed to the Commission's liberal policy initiative, which soon culminated in a long process of negotiated retreats towards an almost complete institutional pluralism by 1995, the British and Norwegian reforms maintained momentum and established new competitive regimes. A more careful analysis of the two reforms reveals, however, that they were distinctly different, both in terms of motivation and structural implications. This juxtaposition highlights the need for a more careful examination of the regulatory dimensions of a liberal electricity regime.

Fig. X.2 presents four essential dimensions of electricity regulation: ownership, concentration, market access and functional decomposition. The two former dimensions represent the structural preconditions. It is traditionally argued that a liberal system requires a fairly decentralised and dominantly privately owned industry. The two later dimensions

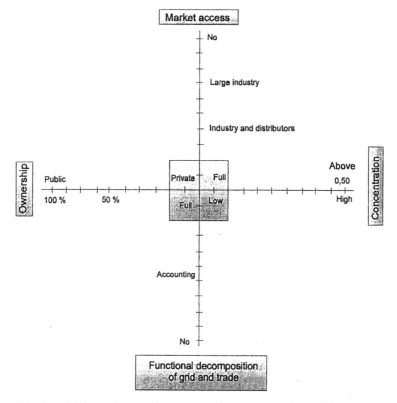

Fig. X.2. Essential dimensions of electricity regulation. *Source:* chapter IV.

refer to regulatory preconditions specific to a grid-bound industry like electricity. It is traditionally argued that a liberal system should provide full market access to all customers, and there should be a complete split between monopolistic and market-exposed functions. Ideally, therefore, a completely liberalised system would be represented by a score on all dimensions close to origo, whereas a centralised public-planned economy system would score on all dimensions close to the periphery.

II.A. The British and Norwegian models

In terms of the above model, the British reform was highly politicised, and concentrated primarily on the ownership dimension, where it implied a massive shift from the public to private shareholders. The Norwegian reform was comparatively far less radical, and basically maintained a tradition of dominant public ownership. It was also more oriented at providing full market access for all consumer segments, down to individual households. Here the British reform only moved gradually, and by the middle of the 1990s still left large consumer segments captives of regional monopolists.

The two reforms also differed dramatically in terms of market concentration. The Norwegian tradition of decentralised municipal ownership had left the country with 129 producers and over 200 distributors. On the distribution side the market was extremely decentralised, with the 20 largest distributors controlling less than a third of the market. With Statkraft, the State-owned production company, owning around one third of the production capacity, the production side was more concentrated. However, with Oslo Energy, the second-largest producer, controlling only 7%, Norwegian Hydro about 6% and the rest of the producers holding market shares below 5%, even the production side remains fairly decentralised.

The British model comparatively figures as much more concentrated. With National Power and Power Gen holding about 60% of the production capacity, the British market figures as highly concentrated (measured by the Herfindal index, used by US competition authorities). On the distribution side, with the 12 regional electricity companies formed out of the previous area boards, the British model is less concentrated.

As far as unbundling, or the degree of functional separation of grid management, trade and production was concerned, the British and Norwegian reforms were quite similar. In both cases, unbundling was only done by accounting, implying that in both countries these activities could be undertaken within the same firm. Fig. X.3 summarises the British and the Norwegian mode of liberalisation.

In terms of regulatory approaches, the British focus on transferring property rights from the state to private owners and on developing a competitive regime on this basis could be seen as reflecting the deep Austrian mistrust of public ownership and public planning. The Norwegian maintenance of public ownership and concentration on a competition policy based on decentralised interaction reveals a stronger confidence in the forcefulness of a structurally based neo-classical competition, where the dynamics of decentralised structural competition is seen as overriding ownership concerns (Midttun and Thomas, 1996).

The gradualistic British approach towards a liberal market follows from its choice of regulation strategy. Dramatic changes to the structure which would make the final shape of the industry less predictable would also reduce the proceeds from privatisation, and would on those grounds be unappealing to the Treasury. Since the Norwegian reform implied no such ownership transfer, the national authorities could

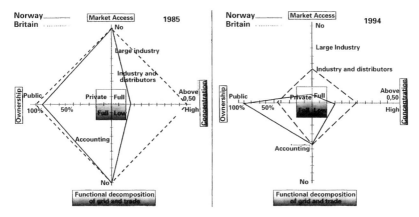

Fig. X.3. British and Norwegian models, 1985 and 1994. Ownership structure and concentration level refers to production. *Source:* Appendix 1.

afford to be less sensitive to the immediate effects of liberalisation on firm value, and open up the market completely to all end-users from day one.

The transfer of ownership in Britain was also more politically controversial (both at the national and company level) than in Norway, where both ownership and structure remained more intact. These differences can also be seen to reflect differences in political culture. In Britain, political decision-making seldom occurs on the basis of consensus-building, and for controversial measures such as the electricity market reform, the usual tactic would to be complete the changes within the span of a single Parliamentary term and to make the changes so radical that undoing them was out of the question. By contrast, the Scandinavian tradition of consensus-building implies a more negotiated and pragmatic approach to industrial policy, ensuring that there was little risk that the changes would be undone simply because of political dogma.

II.B. The strong European position

Along with Britain and Norway, the European Commission was also one of the early forerunners of liberal reform. As shown in chapter IX, the goal was to include these energy sectors under the original European Community vision of 'establishing a common market and progressively approximating the economic policies of Member States ...' (Treaty of Rome, Article 2), a new policy which definitely broke with many of the traditional 'axioms' of the old organisation of the electricity industry.

As already mentioned, the policy outlined by the EC Commission was in many ways a strong and confident liberal policy, reflecting the apparent success of the general 1985 White Paper on the internal market. The first step towards an implementation of the internal energy market was taken by the Commission's Directive proposals in 1989 on price transparency, transit rights and collaboration on infrastructure incentives. In October 1991 the Directives were followed by a set of guidelines for completion of the internal market in gas and electricity.

The Commission continued the pursuit of its strong policy by tabling two proposals for Directives on 22 January, 1992, designed to abolish exclusive rights for the production of electricity and the construction of power lines and gas conduits. The two proposed Directives also called for unbundling, as in the British and Norwegian cases, by separation of management and accounting with respect to production and distribution activities in vertically integrated companies. In line with the British model, the Commission called for full grid access or so-called 'third-party access' for certain distributors, and access for large consumers to their networks.

These two 1992 Directives marked the culmination of the EU's strong policy on the internal energy market, ironically as the general internal market was being put into practice. In November 1992 the Council of Ministers decided that it could not accept the measures as presented, especially third-party access, and asked the Commission to modify them.

From its strong liberal policy position, the Commission had envisaged a European development along the lines of the British and Norwegian reforms (Fig. X.4). It anticipated the Norwegian model in so far as it did not propose ownership transfer, but concentrated on the commercial pressure of market forces under free trade to override political considerations which might be embedded in public ownership. In terms of market access, the Commission's model appeared more 'British', as transit rights were firstly only to be allotted to large companies, but then gradually extended to other parties. As far as unbundling was concerned, the Commission's ambitions coincided with the British and Norwegian demand for separate accounting of the grid and production/trade activities.

II.C. The second wave of Nordic liberalisation: Finland and Sweden

As described in chapter IV, the pioneering liberalisation in Britain and Norway was eventually followed by two more Nordic countries, Finland in 1995 and Sweden in 1996. The Swedish and Finnish liberalisa-

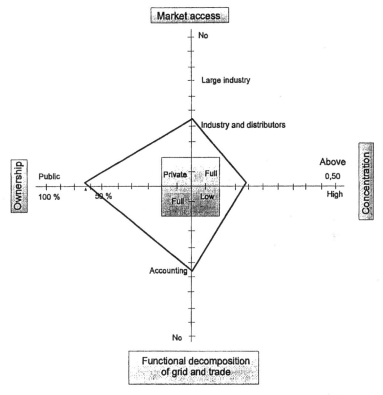

Fig. X.4. The strong EU position for a liberal European regime adapted to the north-western European regime. Ownership and concentration scores are related to sample of eight countries, and refer to electricity production. *Source:* Appendix 2.

tion was carefully co-ordinated with the Norwegian, and by 1996, there-fore, the Nordic region, minus Denmark, was putting into practice the European Commission's idea of a common internal electricity market, although with some transitory limitations for Finland.

Both the Finnish and Swedish reforms followed the Norwegian model in keeping ownership relations intact, and relied on competitive forces in a decentralised market to realise free-trade solutions in spite of a dominant public ownership. This was done in spite of the dominant role of the state company in Sweden, Vattenfall, which commands more than 50% of Swedish production capacity. The concentration of the Finnish electricity industry takes a middle position between the rather decentralised Norwegian position and the highly centralised Swedish position.

One of the main arguments for keeping Swedish Vattenfall intact was that it would operate in a larger Nordic market, where it would not immediately jeopardise free trade. Finland and Sweden did not, however, follow the Norwegian model in opening up for full consumer participation straight away. Both countries here went for a modified British model, where households were kept 'captive' during the first phase of reform.

As far as unbundling was concerned, Finland and Sweden chose different paths. At the central grid level, Sweden followed the Norwegian path and separated out an independent State-owned grid company. Finland, however, followed a modified British model, where a new central grid company is being formed with the electricity industry itself holding the shares. At the regional and local level, Finland chose to follow the British–Norwegian model in only demanding separate accounts for grid operation and other activities, whereas Sweden decided to demand separation through fully separate companies. Fig. X.5 sums up the Finnish and Swedish approaches in terms of our model of the dimensions of electricity regulation.

As shown in chapter IV, the main national grid companies in the emerging Nordic internal market are rapidly developing common institutions and trade regimes to handle inter-Nordic trade. The Norwegian Bourse has, under the new name Norpool, since May 1996 become a joint venture between the Norwegian and Swedish main grid companies, at the same time as being extended to handle Norwegian/Swedish trade. The grid tariffs in the two countries are organised according to the same principles, and the border fee for electricity trade between the two countries has been abolished. At the same time, the Bourse is devel-

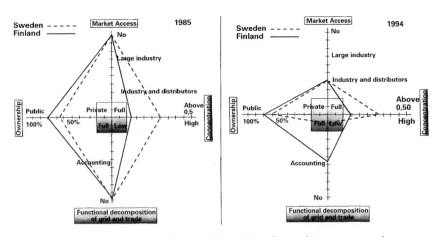

Fig. X.5. Finnish and Swedish models in 1985 and 1996. Ownership structure and concentration levels refer to production. *Source:* Appendix 2.

oping its trading instruments from physical forward exchange to financial futures exchange in addition to the traditional spot trade. The exchange through the new Bourse system comes in addition to a flourishing Nordic bilateral trade. Finland is currently developing its own Bourse for internal Finnish trade, but as the reorganisation of the Finnish grid company takes place, and as new interconnecting capacity is built from Finland, Finland will also join fully in the Nordic market.

The limitations in the Nordic market are now more on the ownership side, where particularly, Norwegian concession rules pose very special restrictions on ownership transfer, and tend to lock companies into traditional strategic patterns. The Swedish and Finnish companies are not exposed to such rules. Harmonisation also remains to be implemented on the taxation side, where the Nordic countries now practice widely differing regimes.

III. The Public-Service Oriented and Cartelised Monopolies

In spite of the liberalist momentum, some countries have retained their support for the planned-economy approach within a monopolistic context. The typical planned-economy approach to this sector in Europe has been the public-service model characterised by public ownership and governed under public law. Germany has also retained an electricity governance regime based on monopoly, however, with a regional anchoring and with large private ownership involved. The German monopoly also stands out from the public-service model in so far as it is anchored in private law through so-called demarcation treaties, and not primarily under public law.

Electricité de France (EDF), the largest European electricity company, has been the dominant proponent of the public-service model, which still enjoys support from the majority of west European countries, particularly in the south. EDF and France have also taken a conceptual leadership in revising this model for application in the new European context. The European Commission has been forced to take this model into account, in the form of the so-called single buyer–single supplier model, following France's successful opposition to the single-market scheme. Compared to France, Germany has remained more defensive.

III.A. The French public-service model

In terms of regulation theory, the French public-service model rests on the 'classical' European post-war model. The basis is the natural monopoly argument, coupled to a goal of egalitarian nation-building which defines the provision of the infrastructure as a national or at least public task. However, while maintaining formal hierarchical organisation as the baseline model, this model is also being reformed to increase performance and efficiency. To achieve this, the public-service model adopts what we have previously (in chapter II) labelled the refinement of 'internal' incentives.

As described in chapter VII, the French electricity system has more or less been governed by State monopoly since World War II. The public-service model has had a strong societal and national anchoring, and EDF's performance in electricity supply is generally regarded as successful. It was under the leadership of EDF that industry and the State succeeded in 'Frenchizing' an American reactor design, and thus became technologically independent. It was also the State company which managed a massive investment and building program based on relatively cheap reactors, without major faults. In the late 1980s and early 1990s, EDF has also succeeded in reducing its dangerously high level of foreign indebtedness.

There is therefore no major economic reason which would justify a shake-up of the existing French model. Nor did there in France, by the mid-1990s, exist a strong ground-swell for a large privatisation and liberalisation of the electricity sector. As argued in chapter VII, French culture leans towards the primacy of collective rules over individual achievements. The change of EDF's status as a public utility would also have to overcome the strong resistance of trade union organisations, which are strongly wedded to the concept of 'public service' and are defending one of its last strongholds within the economy.

The maintenance of a public-service approach has, however, not prevented EDF from innovation. At the commercial level, it has been able to propose special deals for large industrial customers. EDF has also increased the variety of its tariffs. Different cut-off options during peak-load hours enable EDF to propose lower rates in exchange. At the technical level, EDF is improving the quality of its electricity supply. Internal reorganisation and efficiency programs are also being implemented to increase productivity and market orientation. EDF has, furthermore, embarked on a diversification strategy, where the company is engaging in waste management and cable television, and is also continuing its expansion abroad, where it is actively engaged both in Europe, Latin America and Africa.

However, while EDF's public monopoly for its core activities seems stable, at least in the short- and middle-term perspective, there are commercial and institutional indications of a more liberal development in the periphery of EDF's activity sphere in the longer run, paradoxically in part also triggered by EDF's own expansive behaviour. Private companies like Compagnie General des Eaux and Société Lyonnaise des Eaux are infringing on EDF's electricity monopoly in niches, mainly in co-generation. The main strategy of these companies is to enlarge their core activities (water, waste, energy) in complementary markets in order to generate economies of scope in the global management of multiple local public services. These firms are pushing for an increased penetration of new rules of regulation in the energy market.

EDF's response has partly been to hit back at their core markets, but here it is running into obstacles due to its public monopoly situation in electricity. After accusations of unfair competition, the French Ministry of Industry has decided to limit the fields of EDF's diversification. There are also signals of a stronger ministerial interference with electricity regulation, a sign that EDF may be losing some of its hegemony over national decision-making.

III.B. German cartelisation

As already mentioned, Germany shares with France a commitment to the monopoly-supply model, although applied at a regional and not national level. As noted in chapter VIII, the German electricity industry accounts for some of the most powerful companies in the country, and they constitute 'an economic and political power cartel, which until today was able to avoid all attempts at change'. The most influential part of the electricity industry are the nine Verbundschaften, which are interconnected through capital links, that also link them to powerful banking, industrial and regional ownership interests. The regional level is served by regional energy supply companies, most of which are directly or indirectly owned or strongly influenced by the big utilities. The local level is served by municipal utilities, where the big utilities also have some shares.

The regulatory regime in Germany is based on a combination of public and private law. The Law for Promotion of Electricity of 1935 provides the framework conditions and orients the electricity supply towards economic efficiency and safety. In addition, the electricity industry is organised on the basis of demarcation treaties between suppliers and local authorities. These treaties secure the utilities' monopolistic rights within an area on a private-law basis, and thereby violate the spirit of the general competition law. The law against limitation of competition therefore has had to exempt contracts of demarcation from the general prohibition of cartels. Through the so-called 'Stromvertrag', the controversial organisation of the West German electricity sector was transferred to the new Länder.

Since the establishment of the Deregulation Commission in 1988, the cartelised regulation of the German industry has come under increased attack. In 1991, the Commission criticised the German electricity system for charging excessive prices, and suggested an abolition of demarcation treaties, the establishment of neutral high-tension grid operation, and a system of tender for local tariff customers.

In the wake of the CO_2 debate and a general concern with the performance of the German regulatory system, the Bundestag has also called for a Government energy-law review. The Federal Ministry of Environment provided an input into this process, which combined liberal and environmental concerns. In 1993 the Ministry argued for a reform consisting of three elements, all classical parts of the liberal repertoire:

1. functional unbundling of generation, transport and supply;
2. organisation of electricity markets by creating a pool; and
3. environmentally oriented market intervention through environmental charges and taxes.

The Federal Ministry of Industry followed up with a draft proposal for electricity liberalisation in the form of an energy-supply law in much the same vein as the Commission and the Ministry of Environment. It proposed classical liberal elements including unbundling and third-party access.

However, the liberal initiatives met with strong opposition from an alliance between the electricity industry and the municipalities. The majority of the SPD-governed Länder also signalled that the energy-supply law would be halted. Government therefore took the issue off the political agenda, awaiting further liberalisation initiatives from the EU.

The alternative to movement in a liberal direction seems to be a move towards a more decentralised public-service model. The German Labour Party, SPD, has had on its agenda a reform of this kind, featuring:

1. a re-communisation of energy supply;
2. a re-definition of energy supply towards energy services;
3. the expansion of public control to investment and licensing, with a least cost planning aim; and
4. a general feed-in right with a price guarantee for combined heat and power and renewables.

Similar elements are also present in the Green Party's agenda. Given the stalemate between the liberalist and decentralised public-service models, the old regulation by cartelisation remains, although with some competition from renewable supply under the electricity feed-in law.

To summarise, the German and French systems by the mid-1990s still remain restricted from liberal market reform (Fig. X.6). The French system still seems firmly wedded to the monopolistic public-service model, and by its size, resource control and centralised organisation seems capable of withstanding the north European liberalisation forces. In fact, by extensive foreign trade and international investment strategies, EDF is taking an active part in liberalisation outside its own home market.

Compared to France, the German regional monopolies have been under stronger challenge from liberal forces. The challenges initiated by the German competition authorities, followed up by ministerial initiatives, had a liberal potential but could not stand up to the combination of economic and municipal vested interests and red–green political forces. Cartelisation, therefore, remains a dominant feature of the German system. However, the large share of private ownership might represent a potential for liberal reform, given the right political conditions.

The German electricity regime might also be decisive for French and broader European development. As long as Germany, Europe's leading economic power, keeps up a monopolistic electricity system, there is probably enough both political and economic commitment to non-liberal solutions to save the electricity sector from a subsumption under the general liberal European internal market regime. However, should Germany 'tip over' to the liberal side, this might have decisive repercus-

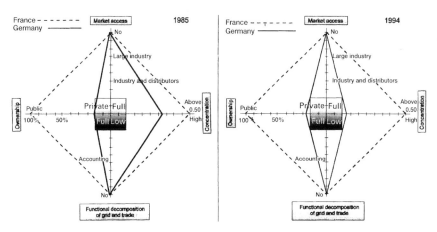

Fig. X.6. The French and German models, 1985 and 1994. Ownership structure and concentration level refer to production. *Source:* Appendix 2.

sions for the general European balance between competitive and monopolistic regimes.

IV. Systems in Negotiated Transition

Both the Dutch and the Danish systems are, by the mid-1990s, in transition towards a liberal model, but gradually and under a negotiated procedure involving the major stake-holders. In terms of institutional and resource-based preconditions, the Dutch mode of electricity regulation bears a strong resemblance to the Danish mode. Both systems have a tradition of decentralised electricity companies which has been kept up until recently, and they have both gained access to vast national gas resources.

In both countries, access to gas supply and locally developed combined heat and power has gradually introduced extensive competition to the traditional electricity-production industry. Through current Government initiatives, both countries may be taking the next step towards internationally oriented wholesale competition.

IV.A. Dutch negotiated reorganisation

As described in chapter VI, the Dutch electricity system is in transition towards a liberal model, but not through a consistent liberal regulatory design, such as in Britain, Norway, Sweden or Finland, but through a process of legislation combined with negotiated agreements between industry and Government.

The regulatory reform process, triggered by the Electricity Act of 1989, which facilitated new opportunities for wholesale trade, set a new basis for distributor–producer relations as well as government–industry negotiations. In several respects, this Act governed the voluntary process of restructuring by the sector itself. A concentration process within the distribution industry has been a major factor determining the strategic consequences of reform. The big provincial distributors have offered very attractive takeover prices to the small municipal owners, which many of them have gladly accepted. In conjunction with the opportunity for liberal wholesale trading within the domestic market, this has turned the distributors into the dominant actor in the Dutch electricity sector.

The Netherlands now finds itself in an intermediary state, in between plan and market, with negotiating bridges between the two. On the one hand, the union of the four large producers, SEP, is still responsible for national co-ordination of production on a daily basis, and for contracts for imports and exports. SEP also develops joint national plans, to be approved of by the Ministry of Economic Affairs every two years.

On the other hand, the municipal and provincial distributors are developing their own production systems, based on decentral combined heat and power, under an allowance made by the Electricity Act of 1989. This development has been so successful that by 1995, 17% of national electricity production was generated decentrally. The result has been an over-production problem, where the use of the central production capacity is declining, with a subsequent decline in the productivity of capital.

The Dutch electricity sector also adapted to new environmental demands through negotiations. In 1990 the sector signed a general agreement with central governmental authorities to reduce emissions of SO_2, NO_x and CO_2. The reduction was to be achieved through investments in technology and efficiency improvements from the producers, and investments in combined heat and power and energy conservation programmes from the distributors.

As stated in chapter VI, there is still much uncertainty as to where the Dutch reform process will finally end. The institutional configuration in the mid-1990s seems unstable, and the system may revert to a more stable state by integrating the decentral production capacity into the existing system of central planning, and thereby fall back on a modified public-service approach. It may also develop further towards a more fully developed liberal system. An Energy Policy Note published by the Dutch government in early 1996 indicates a preference in the liberal direction.

Preliminary steps in a liberal direction, taken by the 1989 Electricity Act, have introduced openness in the Dutch system, but not enough to achieve a major transition. Such a transition might, however, occur if the new Government policy signals win Parliamentary support and materialise into formal legal steps.

IV.B. The Danish transition

Like the Dutch industry, the Danish electricity industry is decentralised. A specific Danish feature is, however, its extensive reliance on consumer ownership. As described in chapter V, 65% of Danish electricity supply comes from directly consumer-owned companies, and the remaining 35% from municipal companies. The Danish system is a non-propfit system, as the electricity companies are not allowed to have a surplus. Because it primarily serves consumer interests, in chapter V Hvelplund describes it as a 'consumer-profit' system. The public-service aspect of the Danish electricity supply is fairly unique in so far as it does not pay tax, either on profit, on assets or concession fees, and until the mid-1970s the Danish electricity supply functioned on a self-government basis, without much intervention by central public authorities.

With the advent of the oil crisis, the nuclear issue and the introduction of natural gas, a certain degree of central governance was introduced in the Danish system. Since 1977, a new 'Law on Electricity Supply' has added a strong element of centralistic planning to the decentralised system. As opposed to France, this element of centralised planning has not been used to implement nuclear technology (the nuclear issue was taken off the political agenda in 1983), but to promote and vigorously expand the natural gas supply. The 1990s also saw an upsurge in environmental concern, which led to central interference in the electricity system in order to reach centrally established environmental goals.

The combination of centralised and decentralised responsibility and planning makes the Danish model both interesting and intriguing from a regulation-theory point of view. The direct representation of consumer interests does, to a certain extent, substitute formal legal governance, and represents a strong pressure from below for cost control. Compared to other national systems included in this study, and excluding the Nordic hydro-based systems with their obvious comparative advantages, the Danish electricitiy supply system performs remarkably well (see Fig. X.1), and together with The Netherlands, maintains some of the lowest prices in western Europe. The open access to information and the low cost of self-governance provided by consumer ownership may be part of the reason for this.

The centralised planning elements of the Danish system serve to co-ordinate the electricity system within a wider energy and environmental context, where demand-side management, alternative technologies and ecological responsibility are central concerns. The extremely high excise taxes imposed through formal legal means by central authorities puts Danish electricity costs at the top of the European list (Fig. X.7), and indicates that in spite of considerable consumer self-governance, central governance is also strong.

Although the interplay between centralisation and decentralisation in the Danish energy system obviously has a complementary character, it also has created tensions, some of which are leading it beyond the public-service model towards a more competitive regime. The introduction of local combined heat and power, under the shelter of the extensive central excise taxes on electricity, has created tensions between the electricity system and new alternative technologies, threatening to produce an electricity surplus. The right of alternative producers to sell to the grid on favourable terms is, according to the electricity system, creating unfair competition and is a threat to planning.

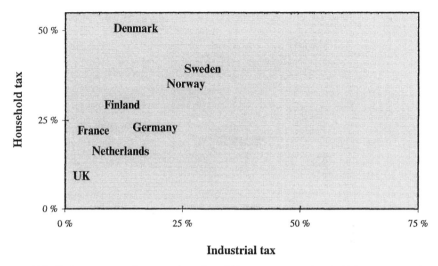

Fig. X.7. Excise taxes on electricity for selected European countries, 1994 (percentage of retail price). *Source:* Appendix 1.

Eventually, the tension within the Danish system seems to be provoking a competitive regime at the wholesale level, and the system seems to be evolving towards a local public-service model within a more liberal national and Nordic system responding to pressure from large customers and neighbouring liberal Nordic regimes.

A recent verdict by the Danish Competition authorities (February 28, 1996) granted third-party access to large customers in the Danish system. Even though this was later over-ruled by Denmark's Competition Appeals Tribunal, it has triggered a swift reform of the Danish Electricity supply law. The new law proposes negotiated third-party access to large industry and distributors (with consumption above 100 GWh), including import rights from foreign countries. At the same time, the law imposes a collective responsibility on all consumers to finance non-competitive environmentally beneficial technologies, such as cheap wind power, through special levies. A passage restricting third-party access with respect to already existing agreements between Danish and foreign companies may, however, indicate considerable limitations to extended Nordic trade.

To summarise, the decentralised character of the Danish and Dutch models has made them both vulnerable to liberal influence, but also capable of a gradual transition to a liberal order. Typical of both countries is a careful gradualistic negotiated transition, where liberal elements are gradually introduced, although the recent verdict by the

Danish competition authorities seems to indicate a transition from negotiations towards a more legalistic model. This is very different from
Norwegian and British liberalisation, which took place suddenly and
under strong central guidance, and through legalistic means.

The strong environmental commitment of both Denmark and The
Netherlands may lead to a stabilisation of wholesale liberalism (Fig.
X.8), leaving the local level monopolistic control over retailing and free
to pursue environmental considerations on a planned or negotiated
basis. The economic efficiency of both regimes under the pre-liberal
regime indicates that the Dutch and Danish electricity industries have
little to fear from European competition.

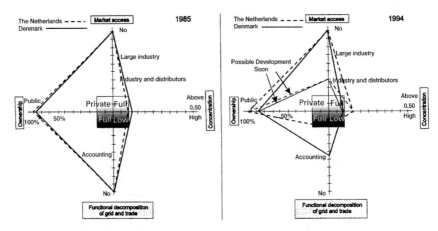

Fig. X.8. The Dutch and Danish Models, 1985 and 1994. Ownership structure and concentration level refer to production. *Source:* Appendix 2.

V. Regulation Under Institutional Pluralism: An Outlook on the Coming European Scene

We have seen from the discussion in chapter II that the regulatory
debate leaves a number of open-ended questions about optimal models
and regulatory efficiency which cannot be answered *a priori*. It is in
principle possible to approach regulatory challenges both by refinement
of internal, organisational and external, market-based incentives. The
former is clearly consistent with a public-service model, and the latter
with a liberal model.

This theoretical conclusion seems compatible with the development
of electricity regulation in the late 1980s and early 1990s, which has left
us with a mixed European scene. The model of public ownership and

monopolistic supply which dominated post-war Europe has been challenged and in part replaced by competitive elements. For four of our eight north-west European cases (Great Britain, Norway, Finland and Sweden), we have seen a definite move in a liberal direction. In addition, The Netherlands and Denmark have taken steps in a liberal direction, although it is still by the mid-1990s, unclear how far the negotiated process is leading.

However, France has strongly maintained a centralised public-service model, and Germany, in spite of liberal reform attempts, remains with a regulatory model based on regionalised cartelisation. At the EU level, the liberal momentum in energy policy has for a long time seemed to be lost, and European authorities have had to embark on an organised retreat from full third-party access *via* negotiated access to institutional pluralism, largely abdicating to the variety of national regimes. In spite of an appearing concensus on a small liberal opening of the EU electricity market, most of the south European countries seem generally wedded to a public-service orientation. Our selection of north-west European cases therefore gives us a pro-liberal bias.

Whereas parts of north-western Europe have left a common stable equilibrium around public ownership and monopolistic supply, it is therefore, by the mid-1990s, rather unclear if, and to what extent, other European countries will follow and where the process will end. Europe is therefore likely to be characterised by institutional pluralism, at least for the coming decade. There are three major reasons for this conclusion. Firstly, the public-service model seems well entrenched, and there are several likely trajectories towards more flexible but non-liberal regulatory regimes, indicating that there may be scope for adaptation to new commercial environments within the public-service framework. Secondly, even the new liberal models span some considerable institutional variation, ranging from oligopolistic capitalist to competitive socialist models. A liberal regime is therefore not itself a well-defined concept, but rather a complex phenomenon with considerable institutional variation, where some variants come fairly close to modified public-service models. Thirdly, it is not easy to find solid north-west European evidence clearly in favour of liberal models (Fig. X.9).

Norwegian electricity prices were extremely low even before market reform, and clearly relate to the country's unique hydro-power resources. Furthermore, they are closely followed by Sweden, which was not liberalised until 1996. The liberal UK price is lower than the public-service based French price, but has increased rather than decreased since liberalisation. Furthermore, the liberal British regime is outdone by the negotiated Dutch and public-service oriented Danish systems in terms of consumer prices. The regionally cartelised German

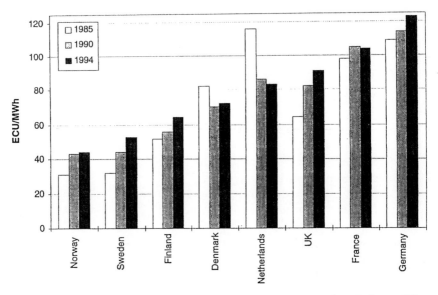

Fig. X.9. Electricity prices exclusive of excise tax in 1985, 1990 and 1995. *Source:* IEA Survey (1994) Q4.

system stands out as the least cost-efficient, but this may reveal hidden taxation through concession fees and coal levies more than industrial inefficiency. There are obviously efficient cost-reducing functions built into the public-service approach, for instance the cost-norm model in the Danish electricity industry, which in spite of Denmark's transition towards liberal trade must be considered a public-service oriented element of the Danish system. Both on the basis of empirical evidence and on the basis of analytical reflection, it is therefore highly arguable that the liberal model should have a clear cost advantage.

In a more cynical perspective, the question of the choice of governance models and regulatory regimes can be seen as a question primarily of distributing surplus among different stake-holder groups. The major relevant groups are consumers, owners, employees and public authorities. The non-profit public-service model is characterised by low profit to the owner, and a reasonable return to employees and consumers (Fig. X.10). This model may or may not be exposed to taxation, which usually is pushed on to consumer prices and reduces consumer utility from the electricity supply. The private-profit based model will tend to transfer the utility from employees to the owners (Fig. X.10) and will usually imply taxation, as an abstention from public taxes usually presupposes a non-profit regime. A regime with large profits

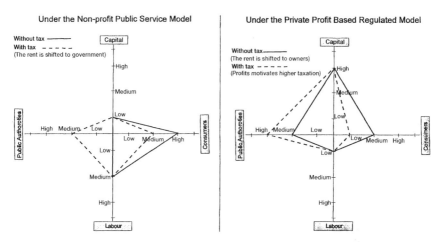

Fig. X.10. Distribution to stake-holders. *Source:* Appendix 2.

to capital could in fact, easily be taken as pretext for taxation increase, with repercussions for consumer prices and direct consumer utility.

Beside the question of stake-holder advantages, the question is of course, whether a shift from a public-service model to a private-profit model will imply productivity gains. Such gains will probably depend on technological development, *vis-à-vis* the scale and scope of organisations. For certain types of innovation, consumer ownership and demand management may be beneficial, whereas for others, private commercial ownership may be more advantageous.

As noted in chapter II, the traditional Austrian–Schumpeterian argument is that competitively exposed privately owned systems will be the most dynamically innovative. However, this perspective tends to neglect the consumer-owned model, as well as the possible trust-creating and organisational capabilities of a pragmatic and even competitively exposed public ownership.

V.A. *Pressure for a new European regulatory style*

In spite of the disputable economic effects and its limited success in reforming the European electricity system, the liberal model has definitely brought new governance elements onto the European scene, not only in the electricity sector, but on a more general scale. These liberal elements, although only partially implemented, are gradually transforming the European style of economic regulation if not by functional supremacy then by institutional isomorphism. As observed by Majone (1993), the new wave of liberal economic and social regulation implies substituting traditional political–bureaucratic governance, often working through publicly owned service production, by means of agencies operating outside the line of hierarchical control or control by the central administration. As we have seen from electricity liberalisation, the establishment of new 'independent' regulation agencies often includes exposure of the service units to competitive pressure.

Our national case studies indicate that this shift may have strong motivation, either politically, commercially, or administratively at the national level. However, the increasing salience of independently governed regulation and competitive markets in Europe clearly also derives from the desire of the EU to play a role in economic governance. Both the tightness and rigidity of the EU budget and the lack of legitimacy *vis-à-vis* nation states precludes EU institutions from pursuing economic governance *via* public ownership, following the classical model of the European post-war state. The Commission's desire to increase its influence by expanding its competencies has therefore had to be channelled through regulation rather than ownership control. The Commission's efforts to establish uniform and liberal economic regulation is supported by multinational firms which prefer dealing with a uniform set of rules rather than with different national regulations. There thus seems to be a structural pressure in support of liberal and supra-national governance both in the process of political and administrative EU integration and in the parallel process of internationalisation of European industry.

V.B. *Regulation and dynamic growth*

The softening of national monopolies and the challenging of functional mono-cultures are also related to the issue of dynamic growth. We have argued, in chapter II, that neither bureaucratic organisation nor regulation based on static optimisation are well suited to handle the challenges of a dynamically evolving economy. When industrial reconfiguration of technology and/or functional sectors is necessary to realise a welfare and/or commercial potential, the regulatory regime must allow for sufficient organisational change and/or strategic adaptation. In terms of regulation theory, the dynamic character of such situations takes us into more of an Austrian–Schumpeterian regulatory world. To this should be added a political dimension of negotiated co-ordination both at the international and sub-national level. It then follows, normatively, that regulation must be 'open-ended' and oriented at flexibly maintaining general fair play and protection against externalities, rather than technology- and institution-specific. Liberalisation, in the form of market exposure, is one way of creating such flexibility. However, opening up for negotiated adaptation and/or allowing for challenges within a public service model may be alternative strategies towards the same end.

It is typical of dynamic adaptation that the strategic initiative lies more in the hands of companies, consumers, regions and municipalities, than with the national state under negotiations within the established EU institutions. The case studies have shown that the established electricity industry within both liberal and public-service models have been challenged by locally organised combined heat and power and renewable technologies, promoted on ecological grounds. What we are here seeing in the public-service oriented systems are challenges between local and central levels of public service. In liberal systems, similar challenges are occurring between producers and distributors.

At the international level, most countries have intensified trade with neighbouring countries far beyond the occasional exchange of surplus capacity which were a part of the old closed, national-systems based approach. Interestingly, public-service oriented France has here been a forerunner, building up an export of 61 Twh (1993) as an outlet for its extensive build-up of nuclear capacity. However, the newly liberalised Nordic countries are also intensifying trade within the internal Nordic energy market, as well as cabling up new connections to continental Europe in order to exploit synergy from matching hydro-based peak-load and coal- and nuclear-based base-load capacities.

V.C. Current challenges

What we see emerging on the international scene is a trade under institutional pluralism, where companies are trading and building strategic alliances across diverse regulatory regimes. Similarly, at the domestic level, we are witnessing innovative strategies which exploit niche opportunities on a scale which often moves the system beyond its traditional boundaries. At both levels, new interplay between companies, consumers and local and regional authorities seem to be getting around the barriers created by the blocked negotiations over common energy regulation at the European level, which by the mid-1990s seem to have been stranded close to a minimum common multiplum.

In the grid part of the electricity industry, the liberalisation of the telecommunications industry has provided opportunities for exploiting electricity grids for multi-functional purposes. We are therefore not only witnessing a realignment of actors, interests and commercial strategies within the electricity industry, but possibly also multi-functional industrial integration. It seems to us, therefore, that the strategic initiative now lies more in the hands of companies, consumers, regions and municipalities than with the national state under negotiations in the established EU institutions although the recent small step towards a common European market regime marks a limited comeback for the political negotiations.

What we may be seeing in Europe for the next decade is a set of multiple processes, where a Nordic liberal *avant-garde* may be creating a 'Schengen' zone of free transaction in liaison with Great Britain, within the context of negotiated relations to a more public-service and cartel-oriented continental setting, perhaps with The Netherlands and Denmark as intermediary zones. Within each of these zones, there are likely to be extensive challenges from local and regional processes as actors experiment with new institutional opportunities and technologies to fit their local needs. This represents a strong challenge to regulators and regulation theory, which must now concern itself not only with dynamic systems, but also with decentralised and multiple systems.

Literature

David, P. A. (1993) Path dependence and predictability in dynamic systems with local network externalities: a paradigm for historical economics. In Foray, Dominique and Freeman Christopher, (eds), *Technology and the Wealth of Nations*, Pinter, London.

Majone, Giandomenico (1993) *Deregulation or Re-Regulation?: Regulatory Reform in Europe and the United States*, Pinter, London.

Midttun, A. and Thomas S. (1996) Theoretical ambiguity and the weight of historical heritage: a comparative study of the British and Norwegian electricity liberalisation, *Energy Policy*, September.

Nugent, N. (1994) *The Government and Politics of the European Union*, London: Macmillan.

EC Energy Monthly 86/17, 26 February 1996, Minister pushes for energy conservation in the EU, London, Financial Times.

EC Energy Monthly 91/24, 12 July 1996, TPA ruling overturned as new power law pending, London, Financial Times.

Appendix 1

Electricity prices in selected European countries, 1980–1994, NOK/kWh

Country	Households				All industry				Heavy industry			
	1994	1990	1985	1980	1994	1990	1985	1980	1994	1990*	1985*	1980*
Norway	0.48	0.46	0.33	0.18	0.23	0.22	0.17	0.07	0.01	0.01	0.09	0.05
Sweden	0.58	0.46	0.30	0.22	0.25	0.26	0.21	0.15	0.24	n/a	n/a	n/a
Denmark	1.31	1.13	1.01	0.64	0.53	0.43	0.54	0.31	0.48	n/a	n/a	n/a
Finland	0.65	0.55	0.46	0.36	0.38	0.34	0.18	0.29	0.33	n/a	n/a	n/a
UK	0.84	0.69	0.54	0.40	0.45	0.39	0.37	0.28	n/a	0.46	0.34	n/a
Germany	1.27	1.16	1.06	0.80	0.66	0.65	0.60	0.46	0.60	0.57	0.43	0.34
Netherlands	n/a	0.84	1.14	0.89	n/a	0.37	0.52	0.46	n/a	0.34	0.29	0.36
France	1.05	1.03	0.99	0.61	0.39	0.39	0.39	0.26	n/a	0.35	0.31	0.25

Sources: Energy prices and taxes, OECD/IEA, Paris, third quarter statistics 1994.
*Imply use of 1991, 1986 and 1982 statistics instead. There is reason to believe that prices to heavy industry are actually lower than the IEA figures, at least for Sweden, Finland and Germany. Prices include VAT and other excise taxes if such exist. All industry is the average of all industrial end-users.

Appendix 2

Country	Market access		Concentration		Functional decompositon		Ownership	
	1985	1994	1985	1994	1985	1994	1985	1994
Norway	No	All	0.0867	0.1153	No	Accounting	90% state	88% state
Sweden	No	All except household	0.3039	0.32	No	Full Decomposition	54% state	59% state
Finland	No	All except household	0.1182	0.1398	No	Accounting	68% state	70% state
Denmark	No	No	0.1297	0.0931	No	Accounting	99% state	89% state
Netherlands	No	No	0.104	0.15	No	Full	100% state	100% state
Germany	No	No	0.3087	0.1276	No	No	NA	27% state
Great Britain	No	Not customers below, 100 kW	1	0.33	No	Accounting	100% state	35% state

Sources: From the case studies in previous chapters.

Chapter XI
The Politics of Electricity Regulation

LUTZ MEZ AND ATLE MIDTTUN

I. Electricity and Politics

Electricity regulation has both the scale and scope to figure prominently in political debate. Firstly, it is a core public infrastructure, upon which both private and industrial users rely heavily. Since the Electricity Supply Industry (ESI) was established as an industry, the role of the utilities has been to provide a secure, cheap and reliable public service. It was indeed seen as a core task for any modernising industrial economy to make electricity available to all end-user segments throughout the nation. The importance of the ESI was clearly demonstrated during the 1973/74 energy crisis, when the sector figured prominently on the political agenda as a crucial societal infrastructure which could not be left to rely heavily on untamed international market forces. Secondly, it is large, including perhaps the largest capital stock ever invested in any one sector. The electric utilities thus administrate immense economic assets, and therefore belong to the most powerful national companies. This economic power position also often translates into political power, and we often find that the ESI has captured a determinant position in the energy policy arena and is frequently able to control its political environment. This power position is based on the one hand on the economic strength of the electricity sector, and on the other hand on structural deficits of the state's energy policy regulation. This means above all a 'state failure' (Jänicke, 1990) in the control of the ESI, and the entanglement of state bodies in the electricity supply industry at all levels. Thirdly, the ESI is an arena for development of and employment of advanced and controversial technology, which naturally makes it an arena for environmental opposition and for local protest. Most particularly, the challenge came from the anti-nuclear movement, which in many countries brought energy issues prominently onto the national political arena and blocked many of the ambitious European nuclear programmes. In a later phase, more general questions of environmental pollution have been at the top of the political agenda.

For these and other reasons, electricity regulation is not only a simple optimisation problem to be left to regulation experts, but also a set of issues with a high symbolic content, large vested interests and considerable political attention. This indicates that the shaping of technological choices and regulatory regimes may be strongly influenced by political forces, and that a politological analysis must therefore supplement pure regulation theory, both in spelling out normative solutions and analysing future developments.

II. Patterns of Electricity Reform Policies

Reform policies have followed very different patterns in the eight countries studied here. The English and Welsh process has been one of politicised polarisation, as opposed to the Nordic reform processes, which have been characterised by a more pragmatic and partly administratively driven market reform. Denmark and the Netherlands have taken a step-wise and negotiated approach, reforming gradually towards a more liberal market order. Germany and France, have remained the laggards in the reform process, however, on the basis of two highly different political configurations. In Germany, close collaborative links between commercial entities and the entanglement of state bodies at all levels, as well as revenue-sharing with German municipalities by way of generous concession fees have, until yet, secured the powerful electricity industry a continued regional monopoly position. In France, the public-service model is well anchored within a strong and reasonably efficient public planning system, which has enjoyed deep societal support, and which is vigorously backed by strong trade unions.

II.A. Electricity reform in England and Wales: politicised polarisation

In path-dependency terms, the English and Welsh (hereafter termed 'British') electricity reform reflects both continuity and disruption. On the one hand, the abrupt policy shift from a public-service to a market-based system conforms with the British tradition of polarised politics. A number of observers have pointed out how the British election system increases governability through strong Parliamentary majorities, but then also lays the country open to strong ideologically motivated policy shifts (Richardson, 1982; Dorfman, 1988). On the other hand, the election of Margaret Thatcher in 1979 and the privatisation program which followed marked a dramatic change, by removing the areas of consensus on the mixed economy between the Conservative and Labour parties, and by including the natural monopolies on the list of issues over which fierce ideological battles could be fought. In the years between the end of World War II and the election of Margaret Thatcher, there was little real conflict about the so-called mixed economy in principle, and particularly not about state control over so-called natural monopolies such as gas, rail, water and electricity. The main area of disagreement between the Labour and Conservative parties was over companies which supplied products and services which were not natural monopolies but which were seen as of strategic importance, such as steel, oil, coal and airlines.

Along with a number of other sectors, privatisation of the ESI and restructuring of the electricity system was launched in the Conservative Party's 1987 election manifesto and in the public debate, with a high ideological profile. The privatisation was implemented with great vigour in a top-down process, where the traditional ESI had very little to say. The argument by the CEGB against its break-up tended to be dismissed as special pleading (Midttun and Thomas, 1996).

The functional autonomy of the British ESI, which had guided energy policy in much of the post-war period, was abruptly set aside by the reinforcement of governmental policy. As such, it was, paradoxically a reassertion of the state, even though the main part of the ESI was privatised and thereby moved out of public hands.

The functional autonomy of the ESI was 'bruised' by the nuclear conflicts in the 1980s (Flam, 1994) and by conflicts within the coal-mining industry, which was a major fuel supplier. Nevertheless, these events had not affected the ESI to the extent that they directly undermined its basic continuation as a technocratic policy arena with considerable elements of capture. The nuclear and local coal conflicts had, however, created enough public disillusionment with the performance of state-owned enterprises to help legitimise Margaret Thatcher's privatisation program.

Even though the privatisation initiative was clearly based on a political–ideological initiative, the Conservative Party's political manifesto did not contain details of the structure to be chosen. As noted by Thomas (chapter III), the act of privatisation itself imposed some limitations on the industrial structuration. Dramatic changes to the structure, which would make the final shape of the industry less predictable, would also reduce the proceeds from privatisation, and would on those grounds be unappealing to the Treasury. Creating a large number of companies would also have been time-consuming and difficult to fit into the available time. Privatisation had to be completed before the next general election, which meant that it could not be planned to take much longer than three or four years.

The specific choice of organisation and regulatory models emerged as a consequence of a complex bargaining game rather than through planning design. Privatisation and the introduction of competition to generation were seen as a way to weaken the power of the British coal industry, which benefitted from the effectively guaranteed market it had for power generation. However, even the Conservative Government had to strike a deal with the coal industry, in so far as an immediate shut-down, which would have been the result of a fully competitive electricity industry, would have been politically unacceptable and would have defeated the Government's plans to privatise the coal industry. In order to allow for a gradual scaling-down of the coal industry through imposing British coal obligations on electricity producers, certain initial monopoly rights had to be secured for the electricity so that implicit coal subsidies could be pushed onto consumers.

The Thatcher Government also had an incentive to modify the competitive pressure of the future commercial electricity regime due to its wish to promote nuclear power. The heavy technological demands and risks the nuclear industry places on the owners implied that private companies would be reluctant to invest in new nuclear power plants. To counter this, the Government had to propose a centralistic structuration of the electricity industry, where a large generation company would inherit two thirds of the CEGB's capacity, which would give it the economic strength to 'shelter' the nuclear plants. To provide an effective counterweight to this large company, all other plant was to be placed in one company.

In spite of these strategies to sugar the nuclear pill, private investors were not willing to swallow it. The nuclear capacity, therefore, had to be separated out into a separate company and financed with heavy subsidies. The two large generation companies, National Power (which would have owned the nuclear power plants) and Power Gen, were retained. This made for a highly centralised electricity market, which again made for an extremely complicated regulatory task.

The British case is therefore a case of a strong top-down state assertion as far as the general direction of the reform policy is concerned, but a case of much more negotiated or even 'muddling through' working-out of the organisational implications and implementation strategy. In this process, the ESI was able to maintain some of its centralised structure, and thereby set the stage for a new market-oriented mode of policy capture, namely through oligopoly games.

II.B. The liberal Nordic systems: pragmatic administrative marketisation

As already noted, three of the Nordic countries (Norway, Finland and Sweden) followed Great Britain in its liberal strategy, but in a more pragmatic, administrative vein, very much in line with the spirit of negotiated political economy which characterises Nordic economic decision-making.

The Norwegian reform was indeed very much administratively driven. Central officials within the Ministry of Finance had for some time been dissatisfied with investments and pricing decisions within the existing electricity regime. Inspired by the British reform and EEC initiatives to liberalise European electricity markets, the Ministry of Finance, in collaboration with the Ministry of Industry and Energy, in 1988 ordered an investigation of a possible Norwegian market reform by the Norwegian School of Economics and Business Administration. The investigation ended up with a recommendation for a decentralised competitive electricity market which was later adopted as part of the reform.

Similar to the Norwegian reform, Sweden's also had solid anchoring within the public administration. During the 1990s, officials in the State Energy Board ('Statens Energiverk') took the initiative to define and describe a market-oriented governance regime for electricity. The developments in Great Britain and Norway, the expectation that a competitively oriented policy would get support within the EEC, and also an increasing appeal among economists for market-based solutions, gave the initial impetus to the reforms, and resulted in a NUTEK[1] report spelling out the principles for Swedish reform. In Sweden, however, the initial stages of the reform were also marked by ideological political commitment. The conservative Bildt-government had a clear party manifesto on liberalisation, and the liberalisation bill drawn up by his Minister of Industry (Westerberg) was clearly part of an explicit liberalist policy. The NUTEK report therefore came in handy as an analytical basis for the reform. However, reform elements were initially announced by the Social Democrats, and the reform has also recently been implemented by a Social Democratic Government.

The Finnish reform was predominantly pragmatically oriented and administratively driven, and seen very much as a necessary adaptation to a coming Nordic and European liberal market. The present supply deficit of about 10% of national consumption has also been a strong motivating factor in the Finnish case. The framework was proposed by

---

[1] As already mentioned in chapter IV, NUTEK is the agency for industrial regulation, including energy

the Social Democratic Government in the early 1990s, but with a governmental change the new Centre Party Government, under President Escko Aho, in coalition with the conservative National Coalition Party, put forward the final proposition.

Whereas the Norwegian reform was decided on and implemented in a one-shot move, the Swedish reform underwent more thorough scrutiny before implementation, which is still pending. This allowed time for learning and functional mobilisation, and large segments of the ESI, have, in alliance with the Social Democrats, taken a more restrictive attitude to reform implementation. Nevertheless, after a longer period of negotiated digesting, the reform passed the Riksdagen with broad support.

The Finnish reform, like the Swedish, was a result of a longer process. Although the reform raised little controversy in party-political terms, there were disputes between electricity consumers and the production industry over its implementation. The protracted reform process was to a large extent the result of successful lobbying for a delay of one to two years by the Finnish electricity industry.

The avoidance of a drastic structural reform such as in Britain probably helped win acceptance for reforms within the ESI, and prevented a strong politicisation along the left–right dimension. Some of the largest electricity companies in fact saw the reforms as opportunities for a larger autonomy and for freer commercial expansion.

The reforms in Norway, Sweden and Finland also gained support from energy-consuming industry, which expected to be able to derive benefits from price cuts. The confederation of industries in both Norway, Sweden and Finland therefore lobbied actively for reform.

The Nordic mode of liberalisation was thus much more a compromise between government policy and traditional ESI values than in Britain. The negotiated character, particularly in Sweden and Finland, along with the fact that the reforms did not interfere strongly with the industrial structure, obviously served to make a compromise possible with the ESI. The reform could in fact also meet demands within major electricity companies, especially the larger state companies, for greater autonomy and freedom from political interference.

II.C. Denmark and the Netherlands: negotiated step-wise transformation

Compared to Great Britain and the liberal Nordic countries, Denmark and the Netherlands have adopted a more gradualistic and negotiated path, somewhere between the market and planning. In both cases, environmental issues were a central policy concern.

The Danish electricity system has been reasonably successful in transforming from a 93% petroleum-based to a dominantly coal-based system since the first oil crisis. This success to a large extent stems from with the high capacity of the political administrative system to work out innovative solutions with other energy-policy stake-holders in a co-operative way. The switching from oil to coal was also combined with saving energy in the heating system by insulation, and use of district heating from power stations.

Since the middle of the 1980s, environmental goals and climate protection have become the core targets of energy policy. 'Energi 2000', the Government's energy policy plan, sets ambitious targets for combined heat and power and wind power. Decentralised heat and power stations are being built at a faster rate than expected, and are expected to constitute around 50% of the heat market. Demand-side management and integrated resource planning are also being introduced.

As noted by Hvelplund in chapter V, the spectacular success of the new environmentally based combined heat and power (CHP) and wind energy supply is presently causing severe tensions within the energy system. The traditional electricity industry is recognising that a stagnation and an imminent over-supply will emerge as result of the large addition of CHP. On this basis, it is protesting against a law proposal which would formalise the right to sell electricity from CHP stations to the grid at long-term marginal costs. However, in some regions, the electricity industry has seen CHP as an investment opportunity, and is now actively involved.

The new boom for local combined heat and power has in many ways introduced a competitive element into the Danish system and the policy conflict marks the conflict of interest between the new local CPH and the old centralised electricity producers. The new development seems to mark a transition to stronger central political–administrative governance, parallel to the strengthening of local municipal authorities and CHP producers. The losing party seems to be the central producers in the old ESI, who are now suffering a serious infringement of their traditional monopoly position.

The strong consensus-seeking tradition in Danish political culture— in part a result of a long history of coalition governments—will probably continue to seek compromises in energy policy. Nevertheless, the new decentralised CHP actors will claim their place at the negotiating table, where they together with large industrial customers will probably gradually push the Danish system into larger openness at the wholesale level.

The Dutch situation is also characterised by a combination of a negotiated simultaneous top-down and bottom-up policy, putting a strain on the production core of the traditional ESI. Although the Dutch system is still in transition, certain features are nevertheless emerging rather clearly. As noted by Arentsen, Künneke and Moll in chapter VI, the opening-up for small-scale co-generation-based electricity production under the Electricity Act, and sponsored by the Government for environmental reasons, proved to trigger a dynamic process where distributors gained considerable independence from, and power over, the traditional production units of the ESI. This again threatened the central planning system with a competitive rather than a co-ordinated development of production capacity. The relative strength of the distributors *vis-à-vis* the traditional producers had of course also been enhanced by the Government's concentration policy, which served to produce larger and more dynamic organisational units.

The contest between central and local elements of the ESI seems to be a logical consequence of the structural opportunities created by the new government legislation, and is in many ways parallel to the consequences of similar opportunities in Denmark. It nevertheless conflicts with the deeply rooted predisposition for consensus-seeking in Dutch political culture (Lijphart, 1968), which might suggest that compromise solutions are likely to be sought rather than letting the actors fight it out on the market arena.

Certain other environmental features in electricity policy, however, seem to conform more closely to the Dutch consensus-seeking norm. The ambitious environmental goals set by Government were largely arranged to be met through negotiated environmental targets and by self-commitments (covenants) of the industry and other actors. In 1992 a covenant with the production companies spelled out reduction figures for SO_2 and NO_x until the year 2000. The distribution companies have also entered into a covenant with the Dutch government targeting a reduction of 17 million tonnes of CO_2 and energy savings of 195 PJ until the year 2000. The close co-operation and trust between Government and industry is also illustrated by the fact that the Dutch ESI administrates subsidies for the Government for the improvement of heat insulation and the introduction of more efficient electrical equipment for households.

The step-wise and negotiated transition pursued by both the Danish and the Dutch seems to mark a third, negotiated way beyond market and plan, where conflicting interests negotiate rights around the bargaining table rather than fight it out openly by economic pressure in the market. The consensus-seeking political cultures of the two countries are probably important preconditions for this development, which may eventually evolve into regulated liberalisation.

III. Alliance Capitalism in Germany

As stated by Mez in chapter VIII, the electricity industry belongs to the most powerful companies in Germany and constitutes an economic and political power cartel, which, until today has been able to resist all attempts to change the framework conditions for energy policy in Germany. The legal and institutional framework cements this structure and secures the privileges of the big utilities.

The powerful ownership links between the electricity industry and major financial and industrial interests in Germany imply that this industry is intimately interwoven into what is termed 'German alliance capitalism' (Shonfield, 1971). As opposed to competitive capitalism, alliance capitalism is characterised by close ties and collaborative relations between commercial entities, and industrial success within this system relies on concerted orchestration of large resources towards common goals. With their huge turnover and monopoly-protected super profits, the national grid companies, which are all interlinked through capital links, are some of the major cash cows of the German economy. The political position of this economic system is consolidated through the entanglement of state bodies at all levels, and through revenue-sharing with German municipalities by way of generous concession fees. We are thus dealing with a relatively clear case of regulatory capture and state failure.

Attempts at reforming the ESI are numerous, and both bottom-up and top-down approaches have failed. In the middle of the 1980s a strategy for a turnaround in energy policy, combined with the re-municipalisation of electricity supply, was spelled out and broadly discussed after the Chernobyl catastrophe. This has remained the policy position of the SPD and the Green parties, also supported by activists at a local level.

Responding to a long-standing critique from the deregulation commission against the monopolistic practices of the electricity industry, the CDU-dominated Federal Government attempted to push through more typical liberal reform elements. A reform proposal, drafted by the Ministry of Industry in October 1993, included partial unbundling, third-party access, and a stricter electricity price control. It was later strongly modified, and finally, in March 1994, withdrawn because of the openly declared resistance of the municipalities, and the signalled opposition from the majority of the SPD-governed Länder in the Bundesrat.

The introduction of environmental concern into the German system has been notably more successful than the liberalisation initiatives. The ordinance on large combustion plants (GFAVO) introduced strict limitations for all emission components such as SO_2, NO_x and particles. With its limitation of private property rights in favour of the environment, the GFAVO constitutes a perfect tool for top-down politics. The technical instructions on air-quality control were supplemented in the same way. The Electricity Feed-in Law, enacted on the initiative of Parliament, provides another remarkable environmentally oriented change in the framework conditions, fundamentally improving the economics of renewable energy. The law guaranteed a minimum payment for electricity from renewable sources supplied to the public grid.

III.A. France: internal reforms within the public-service model

The French ESI is dominantly based on political/administrative governance with the State-owned company, Electricité de France (EDF), in a dominant position both within policy formulation and implementation. This dominant position has in many ways put EDF in a position to influence the electricity policy debate through regulatory capture. It has been able to dominate both the French Ministry of Industry and the French Agency for Environment and Energy Management (ADEME).

As described by Poppe and Couret in chapter VII of this book, the French model must be seen against deep trends in French political culture, such as the strong disposition towards political and economic independence. The struggle to build strong national companies, among them EDF, has been an essential part of breaking the stranglehold which American and other foreign multinationals have on the world energy technology market. The governance of electricity supply through a national monopoly market also follows naturally from the French propensity for central planning.

In a comparative political economy, the French model is frequently classified as 'étatocentristic' capitalism, as compared to British/North American commercial capitalism and German alliance capitalism. In the ideal version of the French model, the State is the sole embodiment of the national interest, and it is the State's role to implement the vision of the 'nation' against private interests and their social and political manifestations.

The French political/administrative orientation is underpinned by trade union interests, which are strongly wedded to the concept of public service and will defend one of their last strongholds within the modern economy. However, French electricity policy also rests on a broad societal consensus, and there are at present no strong national forces pushing for privatisation. EDF's success in diminishing oil dependence and establishing and maintaining a nuclear programme which is unsurpassed in Europe, has given the French ESI a good public standing.

French electricity policy is, however, not without nuances. As described by Poppe and Cauret (chapter VII), private French water companies like the Compagnie Generale des Eaux and Société Lyonnaise des Eaux are infringing marginally on EDF's monopoly domain with local co-generation projects. These companies, which are presently engaged in a wide set of activities (water, waste and energy) are hoping to generate economies of scope by engagement in complementary markets, and are pushing for new rules which will allow them to expand. As far as urban services are concerned, France has developed a long tradition of privatisation through long-term contractual arrangements between municipalities and private operators. The goal of the water companies is to import the French private concession mechanism into the electricity sector. Together with EDF's own engagement in water and cable TV, along with international electricity services, this provides a latent liberalistic undercurrent in French ESI policy. As yet, however, there are no clear indications that the public-service model and EDF's dominant position are severely threatened, and the French policy remains one of adaptation within the public-administrative model, though with some gradual market-opening under strong EU-pressure.

IV. Explaining European Policy Patterns

The institutional diversity of European electricity policy probably precludes any simple explanation and leads us to consider, rather eclectically, a broad theoretical menu. In the following, we will briefly review some of the most relevant politological perspectives, before drawing on them in our explanatory approach. The review includes:

1. American capture theory, interest group theory and agency theory;
2. British 'policy community' perspective;
3. European neo-corporatist tradition;
4. Nordic negotiated political economy traditions;
5. German state discussion;
6. path-dependency theory and;
7. institutional isomorphism theory.

This review serves as a heuristic help to provide a brief overview of the theoretical elements which will later be combined in a more general and eclectic explanatory approach.

IV.A. The American capture theory, interest group theory and agency theory

Due to its prominence on the political arena, electricity or utility regulation has figured prominently in American political-science literature, with a large specialised segment specifically devoted to this topic. According to Gormley (1983), the different explanatory approaches within the American literature can be classified to following categories: capture theory, interest group theory, and agency/administration theory.

Capture theory was developed in the 1950s and 1960s and explains 'administrative decisions as responses to external pressure, exerted primarily and sometimes exclusively by regulated industries' (Gormley, 1982, p. 298). The electricity supply industry expects government to set framework conditions, in which profits can be maximised. In return, the regulated industry provide votes and campaign resources. Capture theory has long historical roots. As early as 1936, Pendleton Herring stressed that regulated industry, under certain conditions, may be able to control public administration and bias administrative decisions to its advantage.

Interest group theory, which emerged in the 1960s and 1970s, characterises administrative decisions as compromises between competing interests and values. This definition includes the regulated industry as one protagonist, taking its place beside a number of actors outside public administration and industry which compete for roles and influence in the regulatory process. Compared to capture theory, interest group theory portrays a more pluralistic and less biased picture. Several studies in this tradition came to the result that the regulating administration is not only controlled by the regulated industry but a number of interest groups which, taken together, tend to level out each other's biases (Salisbury, 1970; Berry, 1984; Crotty *et al.*, 1994). Such interests include industries other than the regulated companies, consumer groups and environmentalists.

The interests group perspective may, however, also be turned towards the administrative apparatus itself. The involvement of multiple agencies with mandates and interests related to various sides of the issue is here seen as a guarantee against biased decisions. Gormley calls this the 'surrogate representation model' (Gormley, 1982, p. 299). In the late 1970s, many state governments set up proxy advocacy offices in order to represent consumer interests in regulatory commission proceedings. Proxy advocates work for the government, but represent the needs of citizens.

Agency/administrative theory is based on the fact that the increasing complexity of the regulatory process and the growing intervention of actors other than the regulatory administration and utilities has led to the situation in which regulators and their staff are playing a more important and influential role in regulatory decision-making. A critical problem is then the ability to ensure that internal decision-making within the public administration follows the general goals laid down for their operation. In terms of principal agency theory, this is treated as a delegation problem, where the subordinate agency (the agent) has greater information than his superior (the principal), who thereby has a control problem (Barney and Ouchi, 1986).

IV.B. The British 'policy community' perspective

The issue of political governance of sectoral industrial complexes with direct implications for electricity policy also figures predominantly in European political science and political sociology, although there are few explicit studies of electricity systems. In Britain, Richardson and Jordan (1979) have analysed sectoral governance using the concept of the 'policy community'. A 'policy community' departs, according to Richardson and Jordan (1979), from the mandate area of a department within the public administration (for instance the electricity department), and is characterised by a committee-based collaboration between organisations on market regulation and governance. In their perspective, co-optation and consensus formation has more influence in determining policy outcomes and market regulation than party policies and parliamentary proclamations.

IV.C. The European neo-corporatist tradition

Compared to the above discussion, the neo-corporatism literature puts greater emphasis on the role of interest organisations and deeper societal and political conflicts when explaining industrial regulation. The main thrust of the neo-corporatist literature has been concerned with negotiated economic governance at the macro level. The neo-corporatist tradition has tended to focus on deeper and more embedded interest conflicts than the American interest group theory. Later works within the neo-corporatist tradition have also showed greater interest for sectoral arenas, such as the ESI. This part of the corporatist literature has moved the perspective down to the meso-level, where the main actors are organisations which represent, regulate and defend sectoral interests (Cawson, 1985; Schmitter, 1986).

The concept of a 'meso-level' relates to the fact that sectoral corporatism finds itself between macro-corporatism on the one side, and a micro-level representing direct negotiations between authorities and the individual enterprise on the other side. Sector governance, or meso-corporatism, is also distinguished from national or macro-corporatist governance by the fact that interest representation is not limited to worker and capital interests, but expanded to include more specialised sectoral organisations (Cawson, 1985).

IV.D. The Nordic negotiated political economy traditions

With strong similarities to the corporatism literature, yet with some distinctive characteristics of its own, the Norwegian study of power, with its focus on negotiated political economy, has been more systems-oriented. It has developed its concept of governance and sectoral regulation through a critique of an idealised liberalist governance model with a focus on the three governance mechanisms: market, democracy and bureaucracy (Hernes, 1978). Somewhat closer to the 'policy community' position, the power study also focused more strongly than the neo-corporatist tradition on the function of the state apparatus in societal regulation.

Nielsen and Pedersen (1989) emphasise the important role of negotiations for economic stability and development in the Nordic countries. They define the negotiated political economy *vis-à-vis* other modes of governance, such as an orthodox market economy and a mixed economy. The negotiated political economy differs from both in the sense that where the orthodox market economist will let the price mechanism play the major role, and where the mixed economist will wish for a clear division between markets and politics, the spokesman for negotiated political economy will suggest negotiations between responsible private, semi-public and public actors.

A central thesis in the Nordic negotiated political economy tradition is that lines of co-operation and conflict in industrial regulation do not run along institutional boundaries like market, politics and administration, as is largely assumed in liberal state theory, but cut right through each of these institutions. The result is a functional fragmentation of societal governance to relatively autonomous sectoral decision-making arenas, one of them being the ESI. In this style of decision making market-based, politically based and administratively based institutions are sectorally oriented in their decision-making, and tend to contradict liberal ideal-type governance by developing extensive *de facto* sectoral autonomy.

IV.E. The German state discussion

A large German literature addresses the regulatory challenge with a central focus on the State. It thereby relates industrial regulation particularly to public institutions. The so-called 'state discussion' (Schuppert, 1989; Hartwich, 1989) in the late 1970s came to focus on the phenomenon of governmental impotence. The term 'state failure' was coined by the political scientist Jänicke (1979, 1990), the economist Recktenwald (1978) and the constitutional lawyer Arnim (1987). This discussion was mainly a response to the inability of the state to deal with environmental problems in industrialised countries, but may in principle also apply to the ESI in general.

The theory of state failure has a clear analogy to the theory of market failure, which is discussed in chapter II of this book. Recktenwald (1978), who firstly diagnosed a 'state failure', concentrated on the 'diseconomies in the state sector', pointing at the wasteful use of tax revenues and the excessive prices of state services. Jänicke (1990, p. 35) takes a broader perspective and distinguishes different forms of state failure:

1. the renunciation of political shaping and preventive intervention towards societal challenges and problems, which he calls 'political failure of the state';
2. the diseconomy of the excessive price of public goods, which he terms 'economic failure of the state'; and
3. the lack of effectiveness of state activity, which he calls 'functional failure of the state'.

IV.F. Path dependency or institutional isomorphism

Although the above-mentioned literature points out a number of aspects which are clearly relevant to the issue, there seems to be quite some way to go to provide a comprehensive general theory which would explain the variations in policy paths taken by the eight European nations included in this study. A more promising path, we think, is to admit the necessity of embedding such an explanation in an institution-specific explanatory approach. Given the large variety of European institutions, we hardly believe that any single general theory will succeed to capture the essential mechanisms, processes and structures in each national case. By instead recognising the institutional and cultural specificities of individual European nations as path-conditioning factors, we are positing policy variation as the natural outcome.

Quite parallel to our critique of the general functional efficiency argument in regulation theory in chapter II, we here argue for a culture and institution-specific modification of the general politological explanations spelled out above. It is hardly necessary here to repeat the path-dependency argument in detail. It should suffice to mention that a central element in the path-dependency perspective is that industrial systems cannot develop independently of previous events (David, 1993), and that local positive loops serve to propagate traditional patterns into future strategic decisions. This implies a development with several equilibrium points, where small events at one point in time may play an important role for future development by determining the course of a long-term development. Transferred to the politics of ESI regulation, path-dependency theory implies that although some of the mechanisms spelled out in the politological literature above may exert strong homogenising pressures, different institutional and political starting points tend to reproduce variety. This variety of regulatory practice may come about because of path-dependent development of the formal institutional structure, but also because of underlying structures and practices, in spite of formal institutional change.

Taking our point of departure in path dependency does not preclude making use of valuable insights from the above-mentioned politological literature. The path-dependency perspective, however, implies that these insights have to be modulated within a specific institutional and cultural framework.

However, the path-dependency perspective must also be balanced against its counterpart, institutional isomorphism. Here we draw on insights arising from the research of Di Maggio and Powell (1983) referring to the constraining processes which lead one unit of a population to resemble another unit because each regime faces the same set of environmental conditions. Di Maggio and Powell (1983) apply this idea to organisations, but in our opinion it may equally well be applied to regulatory regimes.

As noted by Hannan and Freeman (1977) and others, competitive conditions may lead to isomorphism as certain organisational models are selected, for instance, through the operation of market forces. The efficiency hypothesis for the choice of regulatory models may be argued on such grounds. The choice of liberal regulatory regimes for Great Britain, Norway, Sweden and Finland, would thus have to do with the fact that such regimes were more efficient, and that the struggle for international competitiveness would compel (or at least strongly induce) nation states to pursue liberal policies. Contrary this, institutional isomorphism does not assume efficiency as the driving force behind the adoption of new regulatory regimes. Di Maggio and Powell (1983) point out three mechanisms through which institutional isomorphic change occurs: coercive, mimetic and normative isomorphism. Coercive isomorphism stems political influence and the need for political legitimacy, irrespective of efficiency implications, may be a strong motivating force for institutional design. In the context of electricity regulation, the strong formal and informal pressures exerted by the EU Commission for liberalisation in the early 1990s may have been felt as a force or persuasion to undertake national reform. The strong attempts by the EU to create a common legal environment conducive to liberal reform, had they succeeded, might have created a compelling force for liberalisation. Changes in regulatory regimes may also occur as mimetic processes, where changes in relevant reference nations acts as a signal to domestic change, perhaps in response to uncertainty. When regulatory technologies are poorly understood, when goals are ambiguous, or when the environment creates symbolic uncertainty, states may model their regulatory regimes on other states. In this perspective, the regulatory regime of one country tends to be modelled on regimes in other countries which are perceived to be more legitimate or more successful. In addition, isomorphism may be closely associated with normative pressure arising from professionalisation, in those cases where the professions have or adopt strong regime preferences. Professions strive to establish a cognitive base for the legitimisation of their occupational authority, and regulatory shifts may occur as a consequence of new professional orientations and/or changes in dominant professions within a given field. As noted by Perrow (1974), since professionalisation tends to create a pool of almost interchangeable individuals who occupy similar positions across a range of organisations, professional similarities may override variations in tradition and control which might otherwise shape organisational behaviour.

IV.G. Comparative observations

Although relating to some of the same challenges, the electricity policies found in our eight country cases in the late 1980s and early 1990s vary extensively. Britain and three Nordic countries have undertaken liberal reform, although with different levels of politicisation. Denmark and the Netherlands have sought change through gradual, negotiated restructuring. France has primarily developed a policy of internal restructuring of the public service, and German electricity policy has been characterised by several reform attempts but with little reform success, to the effect that both countries remain effectively monopoly-organised.

　　With respect to liberalisation policies, the eight countries included in this study thereby fall into three more or less distinct groups. The liberal pioneers, including Britain and Norway, with Sweden and Finland lagging somewhat behind; the proponents of continued monopolies, including France and Germany; and the gradualistically liberalising countries, including the Netherlands and Denmark.

The grouping together of Britain and Norway in a common liberal forefront illustrates our reservations about a simple explanatory model. If one takes a careful look at the two countries, one can see that they have liberalised for very different reasons. In the British case, explanations can be found both at the general political level and at the industry level. From a political-culture perspective, the polarised conflict between two dominant parties which are often in majority positions creates a political precondition for dramatic policy shifts. The British state ownership of the pre-reform electricity industry could also have served to facilitate electricity reform. Finally, hostile relations between government and the trade unions in the mining industry, which has traditionally been a central fuel-supplier to the electricity industry, implied that there was no strong neo-corporatist consensus policy which would constitute a basis for policy constraint from a government keen on liberalisation.

In terms of both political culture and industrial structure, Norway, the other liberalising pioneer, differs extensively from Britain. The Norwegian political scene has been characterised by minority and coalition governments throughout the 1980s and early 1990s, and does not provide a basis for strong and controversial liberalisation schemes. State ownership is limited to about one third of the electricity industry, which is otherwise dominantly owned by municipalities. The driving forces behind the early Norwegian reforms are therefore highly different from those in Britain. As already mentioned, at the political level the reform may be seen as a reaction to previous centralisation attempts, and not as a result of a strong ideologically motivated policy. Such anti-centralistic sentiments have also been voiced from the municipally dominated industry, wishing to preserve its independent status. The Norwegian reform was also highly facilitated by the existence of a free-trade market among power companies for more than 20 years. The fairly decentralised Norwegian industry, along with dominant public ownership, made for a liberal reform which differed substantially from the British case, and which could be termed 'competitive socialism', thereby also indicating a soft and labour-friendly attitude which would not disturb the traditional Norwegian consensus policy *vis-à-vis* labour.

Sweden and Finland resemble Norway in their pursuit of non-polarised coalition-based industrial policies. However, a larger private-ownership share gives both the Finnish and Swedish industries less of the Norwegian 'municipalistic' flavour. For Sweden, the attachment to the Norwegian (and later Nordic pool) system has been a major facilitator for liberal reform.

To summarise, the three reforming Nordic countries have liberalised their electricity industry in a neo-corporatist/social democratic, non-polarised way. The fairly easy acceptance of free-trade principles may be connected with institutional predisposition, such as a decentralised industrial structure, and the existence of the Norwegian producer-pool with access for Nordic neighbours. It may also, however, be connected with the competitive production costs of the three countries' electricity industry, and the competitive advantage that this would give in a free-trade based north European market. It lies in the pragmatic social-democratic Nordic tradition to have an open-minded attitude to regulatory means, as long as they serve national interests and are compatible with economic stability.

Paradoxically, France and Britain, the two counter-poles in liberalisation policy, share many of the same structural characteristics. They have both traditionally had centralised, state-owned electricity systems, and France resembles Britain in having a fairly polarised political-party structure. However, these similarities take on a fairly different flavour within the two rather different national industrial-policy contexts. The strong French commitment to planning, and to orchestrating industrial policy through a close collaboration between a commercial, administrative and political elite, probably precludes sudden politicised privatisation, as it would violate a basic consensus with wide industrial-policy ramifications. This contrasts with British industrial traditions, where financially based management has a much stronger legitimacy, and where government (with some notable exceptions) does not tend to involve itself as directly in industrial planning. Furthermore, the unions play an important role in French society, and have often proved their ability to mobilise against unpopular government decisions. With its commitment to vulnerable and highly centralised nuclear technology, EDF has formed a pact with its employees to secure stable operation. If the employees were to see this pact threatened, they would not hesitate to take strong action, with disruptive effects for the whole of France's electricity supply.

Apparent structural similarities on important explanatory variables, in spite of opposite policy outcomes, has intrigued us in the British–French comparison. The opposite is the case when we compare France and Germany. Both countries rank as the west European laggards in electricity liberalisation, but they differ extensively both in terms of political and industrial structure. The polarised character of the French party structure does, as already observed, resemble the British situation. This stands in contrast to the German party structure, where both the

Social Democrats (SPD) and the Conservatives (CDU) have to seek alliance partners to achieve majority support in Parliament. Germany and Britain thus conform reasonably well to the political structure/culture hypothesis that drastic reform is most likely to emerge in a context of 'strong politics', based on non-compromising political majorities. France, in this connection, is the exception.

As far as industry structure is concerned, the two liberalisation laggards, France and Germany also differ. Compared to the British case, the private-law based structure of the German industry makes it difficult to direct politically. A dramatic restructuring of privately owned firms largely operating under private law would probably lead to intriguing legal problems and political complexity. This would in turn make dramatic reform by political decree more difficult to implement in Germany than in Britain, where the CEGB used to be wholly state-owned. Again, the German case fits the general hypothesis, while France becomes an exception. However, in a different liberalisation trajectory, where implementation could come about by competition law and not by direct political architecture, the German private-law based system would of course be more adaptable.

The Dutch and Danish cases, where liberal transformation is evolving gradually in a negotiated manner, fit in nicely with what one would expect from the neo-corporatist or negotiated political economy perspectives. The coalition character of the Dutch and Danish Governments, and their consensus-oriented tradition for dealing with labour–capital conflicts, leads them naturally to pursue negotiated and gradualistic energy policy strategies.

The style of negotiation differs somewhat between the two countries. In the Netherlands, Government plays a leading role. By unbundling the electricity industry they have created receptive negotiating partners, and are pursuing liberalisation as well as environmental policies through convenants, or voluntary agreements, but with the threat of an alternative legalistic strategy. In Denmark, the Parliament has been more important than Government in initiating steps towards modernisation and liberalisation of the Danish ESI, through negotiations and resulting voluntary agreements, and to some extent also through legal measures.

IV.H. Elements of an explanatory model

From our comparative observations above, it seems difficult to explain the diversity of European electricity policies within a consistent, general theoretical framework. The problem with general theories is, as we have stated earlier, that each of them taken alone is not capable of capturing

the institutional diversity and probable multiple causality which characterises the variety of European developments. In capture theory, variations in the degree of liberalisation would be explained in terms of
degrees of capture; in 'policy community' theory, by degrees of
co-optation and sector-internal consensus formation; in interest group
theory, in terms of different interest configurations; in corporatist and
negotiated political economy theory, in terms of variations of bargaining
power of relevant actors within the negotiation arena; and in state failure theory, in terms of diseconomies, the lack of effectiveness and political inadequacy of the state.

The complexity of the national developments spelled out above
seems to require a broader, combined understanding, where path
dependency may specify some of the conditioning factors, and selected
general theories may explain some of the moving forces. Furthermore,
our comparative observations indicate that any sensible analysis of
energy policy must be multi-levelled, and include not only policy at the
energy sector level, but also general national policy predisposition.
French 'étatism', Dutch 'pillarisation', Nordic 'negotiated political economies' British 'polarisation' and German 'alliance capitalism' form general national institutional predispositions which also influence sectoral
decisions.

The British shift from a public-service to a market-based model,
which in sectoral terms is a clear path-discontinuity, nevertheless conforms with a more general British path of polarised politics. Given the
electoral support for Conservative Thatcherism, it was no surprise that
Britain was the first country in Europe to turn its infrastructure sectors
upside-down. German alliance capitalism may also explain why the
current electricity regime survives, in spite of extreme regulatory capture and extremely high consumer prices, and the French étatist predisposition gives a plausible explanation of why French electricity-sector
reforms primarily take the form of internal restructurating of the EDF.

Less understandable in path-dependency terms is, perhaps, the
development of liberal electricity regimes in Norway, Finland and
Sweden. In the spirit of negotiated political economies, we would have
expected these countries to pursue a path closer to the Dutch and
Danish, with negotiated adaptation inside the boundaries of a public-
service economy. The pragmatic character of the Nordic social democratic tradition does, however, also allow a flexible use of market elements in societal governance. Furthermore, the existence (for a few
decades) in Norway (the pioneering Nordic country in electricity marketisation) of a spot market for electricity may help explain why electricity-market reform came so quickly to this country. Given the close
institutional and political ties and the extensive grid connections

between the Nordic countries, the spillover effect, especially to Sweden, is also understandable.

The Swedish, Norwegian and Finnish cases, however, illustrate that institutional change takes place within an international context, where external influences outside the boundaries of the nation state are also part of the game. The Norwegian market reform shows many traces of mimetic adaptation to European and British reform ideas, and the Swedish and Finnish reforms have also taken elements of the Norwegian case as a model. Theoretically, we therefore need to balance the path-dependency perspective with elements of institutional isomorphism where international diffusion of regulatory styles and organisational models constitute a counter-force to national idiosyncrasies. Given a certain proliferation of the marketisation of electricity systems, institutional isomorphism helps explain how liberalisation processes may gain momentum both at the political and enterprise levels.

V. Future Politics of Electricity Regulation

The complexity of regulation policy makes it difficult to predict exact short-term policy developments, as indeed it is difficult to explain the existing policy patterns. The contingency on both sectoral, national and international levels opens up several trajectories. Nevertheless, there seems to be a long-term main trend from stable monopolies over regulated liberal regimes towards regular commercial competition.

Great Britain and the Nordic countries (minus Denmark) seem committed to a market course, although with rather a different content and structure. Unacceptable market outcomes may threaten the British model, and lead to pressure for change under a Labour government. It would, however, prove extremely difficult to roll back private ownership without coming into conflict over basic constitutional issues. Most likely, therefore, one would be speaking of a modification rather than a complete change of the market model.

The Nordic liberal regimes seem more likely to avoid strong political confrontation. Firstly, the far more decentralised market structure will probably exert a modifying pressure on oligopolistic behaviour. Secondly, the dominant public ownership also allows a certain political control over dominant actors if they move too far beyond what is politically feasible. Thirdly, the negotiated political economy, in part because of the two former elements, also sets a tone of self-imposed societal responsibility among the market actors.

Neither the Dutch nor the Danish negotiated regimes seem to have found stable solutions, but are taking important steps in a liberal direction. In the Danish case, the apparent success of local combined heat and power has created tension *vis-à-vis* the central producers, for which a solution has not yet been found. A further expansion along the present trajectory for local combined heat and power will imply fundamental changes in the power system. A new opening for wholesale trade in the Nordic pool system is taking the Danish system yet another step in the liberal direction.

In the Netherlands, the ongoing process of economic restructuring is also leading to regulated competition. With the new opening-up for international wholesale competition, Dutch distributors are gaining direct access to north European producers. The process has had (and will most probably still have) a negotiated character, and regime changes will be outcomes of gradualistic moves rather than of strong top-down policies.

The French policy position seems to be heavily anchored in a hierarchical governance model, dominantly dictated by the EDF. However, there seem to be two modifications to this view. Firstly, there has in recent years been a gradual build-up of regulatory and energy-policy expertise at the government level which provides the seeds of a somewhat more independent policy-making capacity. Secondly, there are commercial challenges from the private water companies towards more competition. Thirdly, there are EDF's extensive international commercial engagements, which may eventually contribute to transforming the company's interests in a commercial direction from within.

With a European Union energy policy in the making, some of the policy premises for future national electricity policies may also be determined at the European level. Earlier, in chapter IX, we described the European policy game as complex multi-level bargaining game across multiple institutional contexts. If co-ordinated, and with sufficient authority or consensus to conduct a concerted strategy, the European Union has potentially the possibility of playing a major role in the reshaping of the European electricity industry and its regulatory environment. The Commission's attempts in the late 1980s and early 1990s to implement a strong liberal policy is an example.

However, the tendency in the European electricity policy debate in the mid-1990s has been towards the other end of the scale, closer to the minimum common multiplum of sovereign, national decision-makers, where the least interested party has the most to say. This setback is probably motivated by the great diversity in resource endowments and industrial-policy traditions, implying that national electricity industries would face international competition with very different capabilities. In the near future, therefore, the European level would seem to exert little authoritative influence over national European electricity policies. The institutional complexity of EU decision-making may, however, potentially provide surprises, for example, if the Commission is able to see general competition rules enforced, or if policies from other infrastructure sectors such as telecommunication (where more progress towards a liberal regime has been made), were transferred to the electricity sector. The expansion of the EU to include Sweden and Finland has also strengthened the liberal faction in EU decision-making, both at political and industry levels. Swedish reflections over a 'Schengen' solution to the electricity sector, where some countries move plurilaterally towards closer co-operation, may also prove to create new opportunities for change.

In the spring of 1996 the emerging EU consensus over a gradual opening-up of the electricity markets, even in France and Germany, may indicate a promising collective strategy. By turning the debate to a question of degree rather than principle, there is more scope for compromise, and in the long run, probably more scope for liberal success.

Literature

Arnim, H. v. (1987) Staatsversagen: Schicksal oder Herausforderung? *Aus Politik und Zeitgeschichte*, Vol. B48.

Barney, Jay B. and Ouchi, William G. (1986) *Organizational Economics*, Jossey-Bass, San Francisco.

Berry, Jeffrey M. (1984) *The Interest Group Society*, Little Brown, Boston.

Cawson, A. (1985). *Organised Interests and the State: Studies in Meso-Corporatism*, Sage, London.

Crotty, W., Schwarts, M. A. and Green, J. (1994) *Representing Interests Groups and Interest Group Representation*, University Press of America, Lanhem.

David, P. A. (1993) Path-dependence and predictability in dyanmic systems with local network externalities: a paradigm for historical economics. In Foray, D. and Freeman, C. (eds) *Technology and the Wealth of Nations*, Pinter, London.

Di Maggio, Paul J. and Powell W. W. (1983) The iron cage revisited: institutional isomorphism and collective rationality in organisation fields. In *The New Institutionalism in Organizational Analysis*, Publisher? Place of publication?

Dorfman, Gerald A. (1988) Politics in Western Europe, Stanford: Hoover Institution Press.

Dutton, David (1991) *British Politics since 1945: the Rise and Fall of Consensus*, Blackwell, Oxford.

Gormley, William (1982) Alternative models of the regulatory process: public utility regulation in the States, *Western Political Quarterly*, Vol. 35, pp. 297–317.

Gormley, William (1983) *The Politics of Public Utility Regulation*, University of Pittsburgh Press, Pittsburgh, PA.

Flam, H. and Baldwin, Richard E., *Enlargement of the European Union: the economic consequences for the Scandinavian countries*, London: Centre for economic policy research.

Hannan, Michel T. and Freeman, John (1977) The population ecology of organisations, *American Journal of Sociology*, Vol. 82, March, pp. 929–64.

Hartwich, Hans-Hermann (1989) *Macht und Ohnmacht politischer Institutionen*, *Tagungsbericht*, Westdeutscher Verlag, Opladen.

Hernes, G. (1978) *Forhandlingsøkonomi og blandingasdminstrasjon*, Universeitetforlaget, Bergen.

Herring, E. Pendleton (1936) *Public administration and the public interest*, New York: McGraw-Hill.

Jänicke, Martin (1990) *State Failure. The Impotence of Politics in Industrial Society*, Polity Press, Cambridge.

Jänicke, Martin (1979) *Wie das Industriesystem von seinen Mißständen profitiert*, Freiburg.

John, B. (1991) *British Policy Today*, Manchester: Manchester University Press, Manchester.

Lijphart, Arend (1968) *The Politics of Accomodation*, University of California Press, Berkeley, CA.

Midttun, A. and Thomas, S. (1996) Theoretical ambiguity and the weight of historical heritage: a comparative study of the British and Norwegian electricity liberalisation, *Energy Policy*, September.

Nielsen, K. and Pedersen, Ove K. (1989) *Forhanlingsøkonomi i Norden*, Charlottenlund, Jurist- og Økonomforbundets.

Perrow, C. (1974) *Complex organisation: a critical essay*, New York, Random House.

Recktenwald, H. C. (1978) Unwirtschaftlichkeit im Staatssektor. Elemente einer Theorie des ökonomischen Staatsversagens. In *Hamburger Jahrbuch für Wirtschafts- und Gesellschaftspolitik*, pp. 155–166, Tübingen.

Richardson, J. (1982) *Policy Styles in Western Europe*, Allen and Unwin, London.

Richardson, J. J. and Jordan, A. G. (1979) *Governing under pressure: the policy process in a Post-parliamentary democracy*, Oxford, Martin Robertson.

Salisbury, Robert H. (1970) *Interest Groups in America*, Harper and Row, New York.

Schmitter, P. C. (1981) Interest mediation and regime governability in contemporary Western Europe and North America. In Berger I. (ed.) *Organizing Interests in Western Europe*.

Schuppert, Gunnar F. (1989) Zur Neubelebung der Staatsdiskussion: Entzauberung des Staates oder 'Bringing the State back in?', *Zeitschrift für Staatslehre, öffentliches Recht und Verfassungsgeschichte*, 28. Band, Heft 1, pp. 91–104.

Shonfield, A. (1971) *North American and Western European Economic Policies: Proceedings of the Conference held by the International Economic Association*, McMillan, London.

Whitley, Richard (1994) *The Social Structuring of Forms of Economic Organization: Firms and Markets in Comparative Perspective*, Manchester Business School, Manchester.

Subject Index